Essential MATLAB for Engineers

Essential MATLAB for Engineers

Edited by Natalie Coffman

www.statesacademicpress.com

States Academic Press,
109 South 5th Street,
Brooklyn, NY 11249, USA

Visit us on the World Wide Web at:
www.statesacademicpress.com

ISBN: 978-1-63989-692-9

Trademark Notice: Registered trademark of products or corporate names are used only for explanation and identification without intent to infringe.

Cataloging-in-publication Data

Essential MATLAB for engineers / edited by Natalie Coffman.
p. cm.
Includes bibliographical references and index.
ISBN 978-1-63989-692-9
1. MATLAB. 2. Engineering--Data processing. 3. Numerical analysis--Data processing.
4. Computer-aided engineering. I. Coffman, Natalie.
TA345.5.M42 E77 2023
510.285 536--dc23

Contents

Permissions

List of Contributors

Index

Preface

The world is advancing at a fast pace like never before. Therefore, the need is to keep up with the latest developments. This book was an idea that came to fruition when the specialists in the area realized the need to coordinate together and document essential themes in the subject. That's when I was requested to be the editor. Editing this book has been an honour as it brings together diverse authors researching on different streams of the field. The book collates essential materials contributed by veterans in the area which can be utilized by students and researchers alike.

MATLAB is a matrix programming language primarily designed for scientists and engineers. It is a renowned fourth generation programming language, which has various applications including data science, machine learning and deep learning. It is important for mathematical design, optimization, calculation and analysis. It is also helpful in providing results with precision, speed and accuracy. MATLAB is used extensively for visualization of mathematical calculations and system analysis. The primary applications of MATLAB include deep learning, analysis and visualization of data, and developing a graphical user interface (GUI) and application programming interface (API). It also plays an important role in the computations of numerical linear algebra, simulation, machine learning, data science, and development of algorithms. From theories to research to engineering applications, studies related to all contemporary topics of relevance to MATLAB have been included in this book. It will serve as a reference to a broad spectrum of readers.

Each chapter is a sole-standing publication that reflects each author's interpretation. Thus, the book displays a multi-facetted picture of our current understanding of application, resources and aspects of the field. I would like to thank the contributors of this book and my family for their endless support.

Editor

Numerical Inverse Laplace Transforms for Electrical Engineering Simulation

Lubomír Brančík
Brno University of Technology
Czech Republic

1. Introduction

Numerical inverse Laplace transform (NILT) methods are widely used in various scientific areas, especially for a solution of respective differential equations. In field of an electrical engineering many various approaches have been considered so far, but mostly for a single variable (1D NILT), see at least (Brančík, 1999, 2007b; Cohen, 2007; Valsa & Brančík, 1998; Wu at al., 2001) from plenty of papers. Much less attention was paid to multidimensional variable (nD NILT) methods, see e.g. (Hwang at al., 1983; Singhal at al., 1975), useful rather for more complicated electromagnetic systems. The 2D NILT methods, see e.g. (Brančík, 2005, 2007a, 2007b; Hwang & Lu, 1999), can be applied for a transmission line analysis, or nD NILT methods, $n \geq 2$, for a nonlinear circuits analysis, if relevant Laplace transforms are developed through a Volterra series expansion, see e.g. (Brančík, 2010a, 2010b, Karmakar, 1980; Schetzen, 2006), to highlight at least a few applications. This paper is focused on the class of NILT methods based on complex Fourier series approximation, their error analysis, their effective algorithms development in a Matlab language, and after all, on their selected applications in field of electrical engineering to show practical usefulness of the algorithms.

2. Multidimensional numerical inverse Laplace transform

An n-dimensional Laplace transform of a real function $f(t)$, with $t = (t_1,...,t_n)$ as a row vector of n real variables, is defined as (Hwang at al., 1983)

$$F(s) = \int_0^\infty \cdots_{n\text{-}fold} \int_0^\infty f(t)\exp(-st^T)\prod_{i=1}^n dt_i \,, \tag{1}$$

where $s = (s_1,...,s_n)$ and T means a transposition. Under an assumption $|f(t)| < M\exp(\alpha t^T)$, with M real positive and $\alpha = (\alpha_1,...,\alpha_n)$ being a minimal abscissa of convergence, and the nD Laplace transform $F(s)$ defined on a region $\{s \in C^n: \text{Re}[s] > \alpha\}$, with $c = (c_1,...,c_n)$ as an abscissa of convergence, and the inequality taken componentwise, the original function is given by an n-fold Bromwich integral

$$f(t) = \frac{1}{(2\pi j)^n} \int_{c_1-j\infty}^{c_1+j\infty} \cdots \int_{c_n-j\infty}^{c_n+j\infty} F(s)\exp(st^T)\prod_{i=1}^n dt_i \,. \tag{2}$$

In the papers (Brančík, 2007a, 2007b, 2010b), it was shown for the 1D, 2D, and 3D cases, the rectangular method of a numerical integration leads to an approximate formula whose a relative error is adjustable, and corresponds to the complex Fourier series approximation of a respective dimension. The method has been generalized for an arbitrary dimension n in the recent work (Brančík, 2011).

2.1 Complex Fourier series approximation and limiting relative error

Substituting $s_i = c_i + j\omega_i$ into (2), and using a rectangular rule of the integration, namely $\omega_i = m_i\Omega_i$, and $\Omega_i = 2\pi/\tau_i$ as generalized frequency steps, with τ_i forming a region of the solution $t \in [0,\tau_1) \times ... \times [0,\tau_n)$, an approximate formula is

$$\tilde{f}(t) = \exp(ct^T)\left(\prod_{i=1}^{n}\tau_i^{-1}\right)\sum_{m_1=-\infty}^{\infty}\cdots\sum_{m_n=-\infty}^{\infty}F(s)\exp\left(j\sum_{i=1}^{n}m_i\Omega_i t_i\right), \qquad (3)$$

with $s_i = c_i + jm_i\Omega_i$, $\forall i$. As is shown in (Brančík, 2011), a limiting relative error δ_M of (3) can be controlled by setting $\mathbf{c} = (c_1,...,c_n)$, defining paths of the integration in (2), namely

$$c_i = \alpha_i - \frac{1}{\tau_i}\ln\left(1 - \frac{1}{\sqrt[n]{1+\delta_M}}\right) \approx \alpha_i - \frac{1}{\tau_i}\ln\frac{\delta_M}{n}, \qquad (4)$$

for $i = 1,...,n$, and while keeping the equalities $\tau_1(c_1 - \alpha_1) = ... = \tau_n(c_n - \alpha_n)$. The simplification in (4) is enabled due to small values δ_M considered in practice. The last equation is used for setting up parameters of the nD NILT method relating them to a limiting relative error δ_M required for practical computations.

2.2 Practical computational methods

It should be highlighted that the formula (4) is valid, and a relative error supposed is really achievable by the nD NILT (3), if infinite numbers of terms are used in the series. In practice, it cannot sure be fulfilled, but a suitable technique for accelerating a convergence of infinite series is usable, as is e.g. a quotient-difference (q-d) algorithm (Rutishauser, 1957). Besides, as has been already successfully used for cases of $n \leq 3$, the formula (3) can be rearranged to enable using FFT & IFFT algorithms for an effective computation.

2.2.1 Partial ILTs evaluation technique

The technique of practical evaluation of the n-fold infinite sum (3) follows the properties of the n-fold Bromwich integral (2), namely we can rearrange it into the form

$$f(t_1,t_2,...,t_n) = \frac{1}{2\pi j}\int_{c_1-j\infty}^{c_1+j\infty}\left(\frac{1}{2\pi j}\int_{c_2-j\infty}^{c_2+j\infty}\left(\cdots\frac{1}{2\pi j}\int_{c_n-j\infty}^{c_n+j\infty}F(s_1,s_2,...,s_n)e^{s_n t_n}ds_n\cdots\right)e^{s_2 t_2}ds_2\right)e^{s_1 t_1}ds_1, \quad (5)$$

or shortly

$$f(t_1,t_2,...,t_n) = \mathbb{L}_1^{-1}\left[\mathbb{L}_2^{-1}\left[\cdots\mathbb{L}_n^{-1}\left[F(s_1,s_2,...,s_n)\right]\cdots\right]\right]. \qquad (6)$$

Although the order of the integration may be arbitrary on principle, here the above one will be used for an explanation. Similarly, (3) can be rewritten as

$$\tilde{f}(t_1,t_2,\ldots,t_n) = \frac{e^{c_1 t_1}}{\tau_1} \sum_{m_1=-\infty}^{\infty} \left(\frac{e^{c_2 t_2}}{\tau_2} \sum_{m_2=-\infty}^{\infty} \left(\cdots \frac{e^{c_n t_n}}{\tau_n} \sum_{m_n=-\infty}^{\infty} F(s_1,s_2,\ldots,s_n) e^{jm_n\Omega_n t_n} \cdots \right) e^{jm_2\Omega_2 t_2} \right) e^{jm_1\Omega_1 t_1} \;, \quad (7)$$

with $s_i = c_i + jm_i\Omega_i$. If we define $F_n \equiv F(s_1,\ldots,s_{n-1},s_n)$ and $F_0 \equiv f(t_1,\ldots,t_{n-1},t_n)$, then n consequential partial inversions are performed as

$$\mathbb{L}_n^{-1}\{F_n\} = F_{n-1}(s_1,\ldots,s_{n-1},t_n)\,,$$
$$\mathbb{L}_{n-1}^{-1}\{F_{n-1}\} = F_{n-2}(s_1,\ldots,t_{n-1},t_n)\,,$$
$$\vdots \qquad\qquad\qquad (8)$$
$$\mathbb{L}_1^{-1}\{F_1\} = f(t_1,\ldots,t_{n-1},t_n)\,.$$

As is obvious we need to use a procedure able to make the inversion of Laplace transforms dependent on another $n-1$ parameters, complex in general. Let us denote arguments in (8) by $p_i = (p_1,\ldots,p_{n-1},p_n)$. Then the ILT of the type

$$F_{i-1}(p_{i-1}) = \mathbb{L}_i^{-1}\{F_i(p_i)\} = \frac{1}{2\pi j} \int_{c_i-j\infty}^{c_i+j\infty} F_i(p_i) e^{s_i t_i} ds_i \qquad (9)$$

can be used n times, $i = n,n-1\ldots,1$, to evaluate (8), with $p_n = (s_1,\ldots,s_{n-1},s_n)$, $p_{n-1} = (s_1,\ldots,s_{n-1},t_n)$,\ldots, $p_1 = (s_1,\ldots,t_{n-1},t_n)$, and $p_0 = (t_1,\ldots,t_{n-1},t_n)$, while $p_j = s_j$ for $j \leq i$, and $p_j = t_j$ otherwise. A further technique is based on demand to find the solution on a whole region of discrete points. Then, taking into account $t_{ik} = kT_i$ in (9), with T_i as the sampling periods in the original domain, we can write an approximate formula

$$\tilde{F}_{i-1}(p_{i-1}) = \frac{e^{c_i kT_i}}{\tau_i} \sum_{m=-\infty}^{\infty} \tilde{F}_i(p_i) e^{j2\pi mkT_i/\tau_i} \;, \qquad (10)$$

$i = n,n-1\ldots,1$, and with $\Omega_i = 2\pi/\tau_i$ substituted. As follows from the error analysis (Brančík, 2011) a relative error is predictable on the region $O_{err} = [0,\tau_1) \times\ldots\times [0,\tau_n)$. For $k = 0,1,\ldots,M_i\text{-}1$, $i = 1,\ldots,n$, a maximum reachable region is $O_{max} = [0,(M_1\text{-}1)T_1] \times\ldots\times [0,(M_n\text{-}1)T_n]$. Thus, to meet the necessary condition $O_{max} \subset O_{err}$, we can set up fittingly $\tau_i = M_iT_i$, $i = 1,\ldots,n$. In practice, a region of the calculation is chosen to be $O_{cal} = [0,t_{1cal}] \times\ldots\times [0,t_{ncal}]$, with $t_{ical} = (M_i/2\text{-}1)T_i$, $i = 1,\ldots,n$, to provide certain margins.

2.2.2 FFT, IFFT, and quotient-difference algorithms utilization
As is shown in (Brančík, 2007a, 2010c), the discretized formula (10) can be evaluated by the FFT and IFFT algorithms, in conjunction with the quotient-difference (q-d) algorithm for accelerating convergence of the residual infinite series, see following procedures.

To explain it in more detail, let us consider an r-th cycle in gaining the original function via (9), i.e. $F_{r-1}(p_{r-1}) = \mathbb{L}_r^{-1}\{F_r(p_r)\}$. For its discretized version (10) we have

$$\tilde{F}_{r-1}(s_1,\ldots,kT_r,\ldots,t_n) = \frac{e^{c_r kT_r}}{\tau_r} \sum_{m=-\infty}^{\infty} \tilde{F}_r\left(s_1,\ldots,c_r + jm\frac{2\pi}{\tau_r},\ldots,t_n\right) e^{j2\pi mkT_r/\tau_r} \;. \qquad (11)$$

The above stated formula can be decomposed and expressed also as

$$\tilde{F}_{r-1}(s_1,\ldots,kT_r,\ldots,t_n) = \frac{e^{c_r kT_r}}{\tau_r}\left[\sum_{m=0}^{M_r-1}\tilde{F}_r^{(-m)}z_{-k}^m + \sum_{m=0}^{\infty}\tilde{G}_r^{(-m)}z_{-k}^m + \sum_{m=0}^{M_r-1}\tilde{F}_r^{(m)}z_k^m + \sum_{m=0}^{\infty}\tilde{G}_r^{(m)}z_k^m - \tilde{F}_r^{(0)}\right], \quad (12)$$

where individual terms are defined as

$$
\begin{aligned}
M_r &= 2^{K_r}, \ K_r \text{ integer}, \\
\tilde{F}_r^{(\pm m)} &= \tilde{F}_r(s_1,\ldots,c_r \pm jm2\pi/\tau_r,\ldots,t_n), \\
\tilde{G}_r^{(\pm m)} &= \tilde{F}_r^{(\pm M_r, \pm m)}, \\
z_{\pm k} &= \exp(\pm j2\pi kT_r/\tau_r),
\end{aligned}
\qquad (13)
$$

when $z_{\pm k}^{M_r} = e^{\pm j2\pi k} = 1$, $\forall k$, has been considered, and $\tau_r = M_r T_r$.

As is evident the first and the third finite sum of (12) can be evaluated via the FFT and IFFT algorithms, respectively, while $2P+1$ terms from the infinite sums are used as the input data in the quotient-difference algorithm (Macdonald, 1964; McCabe, 1983; Rutishauser, 1957). We can replace the above infinite power series by a continued fraction as

$$\sum_{m=0}^{\infty}\tilde{G}_r^{(\pm m)}z_{\pm k}^m \approx d_0/(1 + d_1 z_{\pm k}/(1 + \cdots + d_{2P}z_{\pm k})), \ \forall k, \qquad (14)$$

which gives much more accurate result than the original sum truncated on $2P+1$ terms only. The q-d algorithm process can be explained based on a lozenge diagram shown in Fig. 1.

$$
\begin{array}{cccccccc}
e_0^{(0)} & & & & & & & \\
& q_1^{(0)} & & & & & & \\
e_0^{(1)} & & e_1^{(0)} & & & & & \\
& q_1^{(1)} & & q_2^{(0)} & & & & \\
e_0^{(2)} & & e_1^{(1)} & & e_2^{(0)} & & & \\
& q_1^{(2)} & & q_2^{(1)} & & & & \\
e_0^{(3)} & & e_1^{(2)} & & & & & \\
& q_1^{(3)} & & & & & & \\
e_0^{(4)} & & & & & & &
\end{array}
$$

Fig. 1. Quotient-difference algorithm lozenge diagram

The first two columns are formed as

$$
\begin{aligned}
e_0^{(i)} &= 0, & i &= 0,\ldots,2P, \\
q_1^{(i)} &= \tilde{G}_r^{\pm(i+1)}/\tilde{G}_r^{\pm i}, & i &= 0,\ldots,2P-1,
\end{aligned}
\qquad (15)
$$

while the successive columns are given by the rules

$$
\begin{aligned}
e_j^{(i)} &= q_j^{(i+1)} - q_j^{(i)} + e_{j-1}^{(i+1)}, & i &= 0,\cdots,2P-2j, & \text{for} \quad j &= 1,\ldots,P, \\
q_j^{(i)} &= q_{j-1}^{(i+1)}e_{j-1}^{(i+1)}/e_{j-1}^{(i)}, & i &= 0,\cdots,2P-2j-1, & \text{for} \quad j &= 2,\ldots,P.
\end{aligned}
\qquad (16)
$$

Then, the coefficients d_m, $m = 0,...,2P$, in (14) are given by

$$d_0 = \tilde{G}_r^{(0)} , \quad d_{2j-1} = -q_j^{(0)} , \quad d_{2j} = -e_j^{(0)} , \quad j = 1,...,P . \tag{17}$$

For practical computations, however, the recursive formulae stated below are more effective to be used (DeHoog et al., 1982). They are of the forms

$$A_m(z_{\pm k}) = A_{m-1}(z_{\pm k}) + d_m z_{\pm k} A_{m-2}(z_{\pm k}) , \quad B_m(z_{\pm k}) = B_{m-1}(z_{\pm k}) + d_m z_{\pm k} B_{m-2}(z_{\pm k}) , \tag{18}$$

for $m = 1,...,2P$, $\forall k$, with the initial values $A_{-1} = 0$, $B_{-1} = 1$, $A_0 = d_0$, and $B_0 = 1$. Then, instead of the continued fraction (14), we can write

$$\sum_{m=0}^{\infty} \tilde{G}_r^{(\pm m)} z_{\pm k}^m \approx A_{2P}(z_{\pm k})/B_{2P}(z_{\pm k}) , \quad \forall k . \tag{19}$$

The q-d algorithm is a very efficient tool just for a power series convergence acceleration, here enabling (7) to achieve a relative error near its theoretical value defined by (4), see the following examples.

2.3 Matlab listings and experimental errors evaluation
In this part experimental verifications of the nD NILT theory above will first be presented, for one to three dimensional cases, i.e. $n \leq 3$. For such dimensions the Matlab functions have been developed and errors stated on a basis of some sample images with known originals. The Matlab listings of basic versions of the NILT functions are provided, together with examples of their right calling. Another Matlab listings will be discussed in more detail later, in the chapter with practical applications.

2.3.1 One-dimensional NILT
In case of the 1D inverse LT, a well-known Bromwich integral results from (2), namely

$$f(t) = \frac{1}{2\pi j} \int_{c-j\infty}^{c+j\infty} F(s)e^{st}dt , \tag{20}$$

where indexes 1 were omitted. By using the theory above a path of the numerical integration is stated according to (4), leading to

$$c = \alpha - \frac{1}{\tau}\ln\left(1 - \frac{1}{1+\delta_M}\right) = \alpha + \frac{1}{\tau}\ln\left(1 + \frac{1}{\delta_M}\right) \approx \alpha - \frac{1}{\tau}\ln\delta_M . \tag{21}$$

In contrast to most other approaches, the 1D NILT method described here enables to treat complex images resulting in complex originals as no real or imaginary parts are extracted during an evaluation process. It can be useful in some special applications, not only in the electrical engineering. We can show it on a simple transform pair

$$F(s) = \frac{1}{s-j\omega} = \frac{s}{s^2+\omega^2} + j\frac{\omega}{s^2+\omega^2} \quad \mapsto \quad f(t) = e^{j\omega t} = \cos\omega t + j\sin\omega t . \tag{22}$$

Of course, when preprocessing the transform to arrange it to a Cartesian form, as is shown on the right sides in (22), the result could by get by inverting the real and imaginary parts separately, by using an arbitrary NILT method. Here, however, no symbolic manipulations are needed in advance, and $F(s)$ enters the NILT function in its basic form as a whole.

A Matlab language listing is shown in Tab. 1, where the relative error needed is marked by Er and is subject to a change if necessary, similarly as the minimal abscissa of convergence (exponential order) α, alfa, numbers of points for the resultant solution, M, and for the q-d algorithm, P. If only real transforms $F(s)$ are considered the bottom line in the listing can be inactivated. The NILT function is called from a command line as follows: niltc('F',tm); where F is a name of another function in which the $F(s)$ is defined, and tm marks an upper limit of the original variable t. In our case, and for $\omega = 2\pi$, this function can have a form

```
function f=expc(s)
f=1./(s-2*pi*j);
```

```
% ------ 1D NILT for complex arguments - basic version -------
% ----------- based on FFT/IFFT/q-d, by L. Brančík -----------
function [ft,t]=niltc(F,tm);
alfa=0; M=256; P=3; Er=1e-10;                    % adjustable
N=2*M; qd=2*P+1;
t=linspace(0,tm,M); NT=2*tm*N/(N-2); omega=2*pi/NT;
c=alfa+log(1+1/Er)/NT; s=c-i*omega*(0:N+qd-1);
Fs(1,:)=feval(F,s); Fs(2,:)=feval(F,conj(s));
ft(1,:)=fft(Fs(1,1:N)); ft(2,:)=N*ifft(Fs(2,1:N));
ft=ft(:,1:M); D=zeros(2,qd); E=D;
Q=Fs(:,N+2:N+qd)./Fs(:,N+1:N+qd-1);
D(:,1)=Fs(:,N+1); D(:,2)=-Q(:,1);
for r=2:2:qd-1
    w=qd-r;
    E(:,1:w)=Q(:,2:w+1)-Q(:,1:w)+E(:,2:w+1); D(:,r+1)=-E(:,1);
    if r>2
        Q(:,1:w-1)=Q(:,2:w).*E(:,2:w)./E(:,1:w-1);
        D(:,r)=-Q(:,1);
    end
end
A2=zeros(2,M); B2=ones(2,M); A1=repmat(D(:,1),[1,M]); B1=B2;
z1=exp(-i*omega*t); z=[z1;conj(z1)];
for n=2:qd
    Dn=repmat(D(:,n),[1,M]);
    A=A1+Dn.*z.*A2; B=B1+Dn.*z.*B2; A2=A1; B2=B1; A1=A; B1=B;
end
ft=ft+A./B; ft=sum(ft)-Fs(2,1); ft=exp(c*t)/NT.*ft;
ft(1)=2*ft(1);
figure; plot(t,real(ft));
figure; plot(t,imag(ft));                        % optional
```

Table 1. Matlab listing of 1D NILT method accepting complex arguments

As is obvious, the Laplace transform must be defined to enable Matlab array processing, i.e. element-by-element array operators have to be used. Thus, the calling our function can look like niltc('expc',4); if the function is saved under the same name, expc, or it is placed inside the M-file with own NILT function (Tab. 1), following always its body. Alternatively, the calling can look like [ft,t]=niltc('F',tm); if respective variables in the brackets are to be saved in the memory after the function finishes.

Graphical results and corresponding errors are shown in Fig. 2. Because the originals are bounded by values ±1, and $\alpha = 0$, we can see the errors satisfy (21) very well ($\delta_M = 10^{-10}$ was considered), excluding only beginning of the interval.

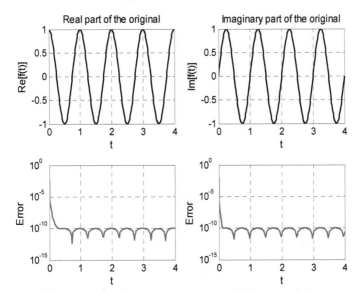

Fig. 2. Numerical inversion leading to complex original $f(t) = \exp(j\omega t)$

Another test functions are considered in Tab. 2, with numerical results shown in Fig. 3. As is again obvious from Fig. 3 the relative errors satisfy theoretical expectations, with an exception of vicinities of discontinuities.

i	1	2	3	4	5	6
$F_i(s)$	$\dfrac{1}{s+1}$	$\dfrac{1}{(s+1)^2}$	$\dfrac{2\pi}{s^2+4\pi^2}$	$\dfrac{1}{\sqrt{s^2+1}}$	$\dfrac{e^{-\sqrt{s}}}{s}$	$\dfrac{e^{-s}}{s}$
$f_i(t)$	e^{-t}	te^{-t}	$\sin(2\pi t)$	$J_0(t)$	$\mathrm{erfc}\left(\dfrac{1}{2\sqrt{t}}\right)$	$1(t-1)$

Table 2. Test Laplace transforms for errors evaluation

2.3.2 Two-dimensional NILT
In case of the 2D inverse LT, a two-fold Bromwich integral results from (2), namely

$$f(t_1,t_2) = -\frac{1}{4\pi^2}\int_{c_1-j\infty}^{c_1+j\infty}\int_{c_2-j\infty}^{c_2+j\infty} F(s_1,s_2)e^{s_1t_1+s_2t_2}\,ds_1ds_2\,, \tag{23}$$

and by using the theory above the paths of numerical integrations are stated based on (4) as

$$c_i = \alpha_i - \frac{1}{\tau_i}\ln\left(1-\frac{1}{\sqrt{1+\delta_M}}\right) \approx \alpha_i - \frac{1}{\tau_i}\ln\frac{\delta_M}{2}, \quad i=1,2\,. \tag{24}$$

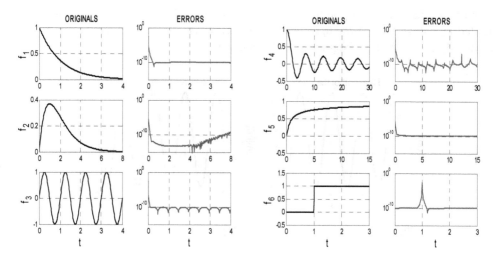

Fig. 3. Computed originals and errors for test Laplace transforms in Tab. 2

A Matlab language listing is shown in Tab. 3, with all the parameters denoted by similar way as in the previous case.

Nevertheless, the Laplace transform variables and the original variables have changed in this listing as $s_1 \to p$, $s_2 \to q$, and $t_1 \to x$, $t_2 \to y$, respectively, which simplified a writing. The parameters are then indexed in compliance with these new notations. Besides, numbers of points used to plot three-dimensional graphical results are set by xpl and ypl.

For the same reasons as at the 1D NILT, the 2D NILT method discussed here enables to treat complex images of two variables resulting in complex originals. We will show it on a simple transform pair

$$F(s_1,s_2) = \frac{1}{(s_1 - j\omega_1)(s_2 - j\omega_2)} \quad \longmapsto \quad f(t_1,t_2) = e^{j(\omega_1 t_1 + \omega_2 t_2)} \ . \tag{25}$$

After rearranging the above equation, we can also write

$$F(s_1,s_2) = \frac{s_1 s_2 - \omega_1 \omega_2}{\left(s_1^2 + \omega_1^2\right)\left(s_2^2 + \omega_2^2\right)} + j\frac{\omega_2 s_1 + \omega_1 s_2}{\left(s_1^2 + \omega_1^2\right)\left(s_2^2 + \omega_2^2\right)} \quad \longmapsto$$

$$f(t_1,t_2) = \cos(\omega_1 t_1 + \omega_2 t_2) + j\sin(\omega_1 t_1 + \omega_2 t_2) \tag{26}$$

The 2D NILT function is called from a command line as follows: nilt2c('F',xm,ym); where F is a name of another function in which the $F(p,q)$ is defined, and xm and ym mark upper limits of the original variables x and y. In our case, and for $\omega_1 = \omega_2 = 2\pi$, this function can have a form

```
function f=exp2c(p,q)
f=1./(p-2*pi*j)./(q-2*pi*j);
```

and its calling can look like nilt2c('exp2c',3,3); with graphical results in Fig. 4. As the originals are bounded by values ±1, and $\alpha_1 = \alpha_2 = 0$, we can see the errors satisfy (21) very well ($\delta_M = 10^{-8}$ was considered), excluding beginnings of the 2D region.

```
% ----- 2D NILT based on partial inversions, by L. Brančík -----
function fxy=nilt2c(F,xm,ym);
alfax=0; alfay=0; Mx=256; My=256; P=3; Er=1e-8;      % adjustable
xpl=64; ypl=64;                                       % adjustable
Nx=2*Mx; Ny=2*My; qd=2*P+1; Ke=log(1-1/sqrt(1+Er));
nx=2*xm*Nx/(Nx-2); ny=2*ym*Ny/(Ny-2);
omegax=2*pi/nx; omegay=2*pi/ny; sigx=alfax-Ke/nx;
sigy=alfay-Ke/ny; qdl=qd-1; Nxw=Nx+qdl; Nyw=Ny+qdl;
Asigx=sigx-i*omegax*(0:Nxw); Asigy=sigy-i*omegay*(0:Nyw);
Asigx2=cat(2,Asigx,conj(Asigx));
rx=[1:Mx/xpl:Mx,Mx]; ry=[1:My/ypl:My,My];
x=linspace(0,xm,Mx); y=linspace(0,ym,My); x=x(rx); y=y(ry);
[q,p]=meshgrid(Asigy,Asigx2); Fpq(:,:,1)=feval(F,p,q);
[q,p]=meshgrid(conj(Asigy),Asigx2); Fpq(:,:,2)=feval(F,p,q);
Fpyp=Pnilt(Fpq,Ny,ry,qd,y,ny,omegay,sigy); % Pnilt to get F(p,y)
Fpy(:,:,1)=Fpyp(1:Nxw+1,:).';
Fpy(:,:,2)=Fpyp(Nxw+2:2*Nxw+2,:).';
fxy=Pnilt(Fpy,Nx,rx,qd,x,nx,omegax,sigx);  % Pnilt to get f(x,y)
figure; mesh(x,y,real(fxy));
figure; mesh(x,y,imag(fxy));                         % optional
% ------ PARTIAL NILT based on FFT/IFFT/Q-D, by L.Brančík ------
function fx=Pnilt(Fq,N,grid,qd,xy,nxy,omega,c);
fx(:,:,1)=fft(Fq(:,:,1),N,2); fx(:,:,2)=N*ifft(Fq(:,:,2),N,2);
fx=fx(:,grid,:); delv=size(Fq,1); delxy=length(xy);
d=zeros(delv,qd,2); e=d; q=Fq(:,N+2:N+qd,:)./Fq(:,N+1:N+qd-1,:);
d(:,1,:)=Fq(:,N+1,:); d(:,2,:)=-q(:,1,:);
for r=2:2:qd-1
    w=qd-r; e(:,1:w,:)=q(:,2:w+1,:)-q(:,1:w,:)+e(:,2:w+1,:);
    d(:,r+1,:)=-e(:,1,:);
    if r>2
        q(:,1:w-1,:)=q(:,2:w,:).*e(:,2:w,:)./e(:,1:w-1,:);
        d(:,r,:)=-q(:,1,:);
    end
end
A2=zeros(delv,delxy,2); B2=ones(delv,delxy,2);
A1=repmat(d(:,1,:),[1,delxy,1]); B1=B2;
z1(1,:,1)=exp(-i*omega*xy); z1(1,:,2)=conj(z1(1,:,1));
z=repmat(z1,[delv,1]);
for n=2:qd
    Dn=repmat(d(:,n,:),[1,delxy,1]);
    A=A1+Dn.*z.*A2; B=B1+Dn.*z.*B2; A2=A1; B2=B1; A1=A; B1=B;
end
fx=fx+A./B; fx=sum(fx,3)-repmat(Fq(:,1),[1,delxy]);
fx=repmat(exp(c*xy)/nxy,[delv,1]).*fx; fx(:,1)=2*fx(:,1);
```

Table 3. Matlab listing of 2D NILT based on partial inversions

Another simple example shows a shifted 2D unit step, with different shifts along the axis. A corresponding transform pair is

$$F(s_1,s_2)=\frac{\exp(-2s_1-s_2)}{s_1s_2} \mapsto f(t_1,t_2)=\mathbf{1}(t_1-2,t_2-1). \tag{27}$$

In this case, a displaying imaginary part gives a zero plane, and the respective line in the 2D NILT function can be inactivated. The graphical results are depicted in Fig. 5, including an absolute error. The respective Matlab function can be of a form

```
function f=step2(p,q)
f=exp(-2*p-q)./p./q;
```

and called as `nilt2c('step2',4,4);`, with the results theoretically expected.

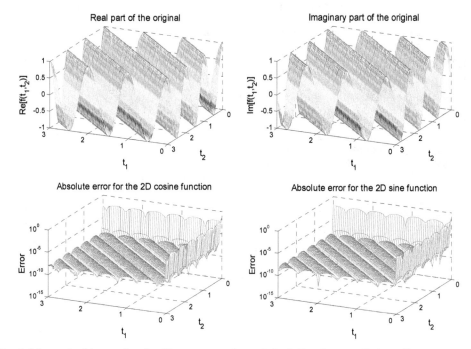

Fig. 4. Numerical inversion leading to complex original $f(t_1,t_2) = \exp(j\omega(t_1+t_2))$

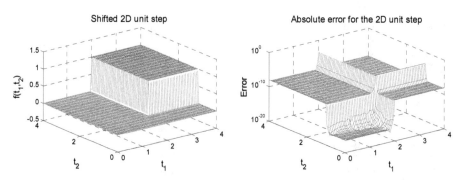

Fig. 5. Numerical inversion leading to shifted 2D unit step $f(t_1,t_2) = \underline{1}(t_1-2,t_2-1)$

2.3.3 Three-dimensional NILT
In case of the 3D inverse LT, a three-fold Bromwich integral results from (2), namely

$$f(t_1,t_2,t_3) = \frac{j}{8\pi^3} \int_{c_1-j\infty}^{c_1+j\infty} \int_{c_2-j\infty}^{c_2+j\infty} \int_{c_3-j\infty}^{c_3+j\infty} F(s_1,s_2,s_3)e^{s_1t_1+s_2t_2+s_3t_3}\,ds_1ds_2ds_3 , \tag{28}$$

and by using the theory above the paths of numerical integrations are stated based on (4) as

$$c_i = \alpha_i - \frac{1}{\tau_i} \ln\left(1 - \frac{1}{\sqrt[3]{1+\delta_M}}\right) \approx \alpha_i - \frac{1}{\tau_i} \ln\frac{\delta_M}{3}, \quad i = 1,2,3.\tag{29}$$

Here only experimental results will be shown to verify an accuracy of the method. A Matlab language listing looks similarly like for the 2D NILT case, but the partial NILT subfunction is called once more, and respective arrays dimensions are enlarged. Original functions corresponding to 3D Laplace transforms cannot be displayed graphically as a whole, of course. However, for one variable chosen as constant, it is posssible to display three respective two-dimensional cuts. It will be demonstrated on the example of 3D shifted unit step, with a Laplace transform pair

$$\frac{\exp(-s_1 - 2s_2 - 3s_3)}{s_1 s_2 s_3} \quad\mapsto\quad \underline{1}(t_1 - 1, t_2 - 2, t_2 - 3),\tag{30}$$

with different values of shifts along respective coordinates so that correctness of results can easily be identified, see Fig. 6. Errors again correspond to theoretically expected ones.

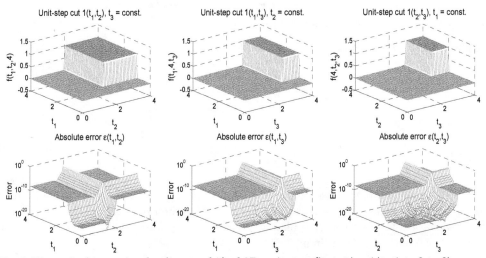

Fig. 6. Numerical inversion leading to shifted 3D unit step $f(t_1,t_2,t_3) = \underline{1}(t_1-1,t_2-2,t_3-3)$

3. Application of NILT algorithms to electrical engineering simulation

In this chapter some examples of the application of the NILT algorithms developed relating to problems of electrical engineering simulation are presented. First, the 1D NILT method is applied for the solution of transient phenomena in linear electrical circuits with both lumped and distributed parameters. This well-known approach is usable wherever linear ordinary differential equations (ODE) are transformed into algebraic ones so that an inverse Laplace transform can be considered. Then the 2D NILT method is utilized to solve transient phenomena on transmission lines (TL) after relevant telegraphic equations (a type of partial differential equations (PDE)) are transformed into algebraic ones by a 2D Laplace transform. In this way voltage and/or current distributions along the TL wires can be determined in a single calculation step. Finally, the utilization of the 1D to 3D NILTs to weakly nonlinear electrical circuits solution is discussed. In this case the relevant nonlinear ODEs describing the circuit are expanded into Volterra series which respective NILTs are applied on.

3.1 One-dimensional NILT algorithm application
3.1.1 Preliminary example based on lumped parameter circuit
A simple example demonstrating the application of the basic 1D NILT algorithm in Tab. 1 is shown in Fig. 7. This really initiatory linear electrical circuit was chosen with an intention to be also considered later, in chapter 3.3.1, as a nonlinear circuit, with G_2 being a nonlinear element. In this way one will be able to compare results and make some conclusions.

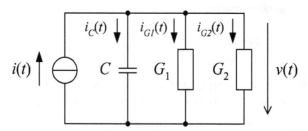

Fig. 7. Linear reactive electrical circuit of the 1st order

Denoting $G = G_1 + G_2$, the 1st-order linear ODE has a form

$$C\frac{dv(t)}{dt} + Gv(t) = i(t),$$

(31)

with a Laplace-domain solution

$$V(s) = \frac{I(s) + Cv(0)}{G + sC},$$

(32)

with an initial condition $v(0)$. Even if the above circuit is very simple a finding time-domain solution could be rather work-intensive if the circuit is excited from some non-trivial input current waveform. A few basic examples are given in Tab. 4, specially the first one results in a transient characteristic of the circuit.

k	1	2	3	4
$i_k(t)$	$I_0 \underline{1}(t)$	$I_0 e^{-5t}\underline{1}(t)$	$I_0 \sin(2\pi t)\underline{1}(t)$	$I_0 \cos(2\pi t)\underline{1}(t)$
$I_k(s)$	$\dfrac{I_0}{s}$	$\dfrac{I_0}{s+5}$	$\dfrac{2\pi I_0}{s^2 + 4\pi^2}$	$\dfrac{sI_0}{s^2 + 4\pi^2}$

Table 4. Exciting current source waveforms and their Laplace transforms

For the above examples, of course, time-domain analytical solutions can be found e.g. based on a Heaviside formula. The 1D NILT function graphical results, under a condition $v(0) = 0$, and considering values $C = 1mF$, $G_1 = G_2 = 10mS$, and $I_0 = 1mA$, are shown in Fig. 8.
The above waveforms can be got by either successive application of a basic version of the 1D NILT method according to Tab. 1, or a generalized 1D NILT function, its vector version, can be used to process all the computations in parallel. This function is shown in Tab. 5.

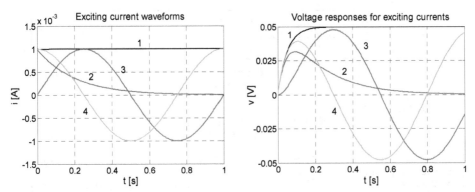

Fig. 8. Numerically computed exciting current and voltage responses waveforms

```
% ------ 1D NILT for complex arguments - vector version ------- %
% ---------- based on FFT/IFFT/q-d, by L. Brančík ----------- %
function [ft,t]=niltcv(F,tm,depict);
alfa=0; M=256; P=3; Er=1e-10;                       % adjustable
N=2*M; qd=2*P+1; t=linspace(0,tm,M); NT=2*tm*N/(N-2);
omega=2*pi/NT;
c=alfa+log(1+1/Er)/NT; s=c-i*omega*(0:N+qd-1);
Fs(:,:,1)=feval(F,s); Fs(:,:,2)=feval(F,conj(s)); lv=size(Fs,1);
ft(:,:,1)=fft(Fs(:,:,1),N,2); ft(:,:,2)=N*ifft(Fs(:,:,2),N,2);
ft=ft(:,1:M,:);
D=zeros(lv,qd,2); E=D; Q=Fs(:,N+2:N+qd,:)./Fs(:,N+1:N+qd-1,:);
D(:,1,:)=Fs(:,N+1,:); D(:,2,:)=-Q(:,1,:);
for r=2:2:qd-1
    w=qd-r;
    E(:,1:w,:)=Q(:,2:w+1,:)-Q(:,1:w,:)+E(:,2:w+1,:);
    D(:,r+1,:)=-E(:,1,:);
    if r>2
    Q(:,1:w-1,:)=Q(:,2:w,:).*E(:,2:w,:)./E(:,1:w-1,:);
    D(:,r,:)=-Q(:,1,:);
    end
end
A2=zeros(lv,M,2); B2=ones(lv,M,2); A1=repmat(D(:,1,:),[1,M,1]);
B1=B2; z1=repmat(exp(-i*omega*t),[lv,1]); z=cat(3,z1,conj(z1));
for n=2:qd
    Dn=repmat(D(:,n,:),[1,M,1]);
    A=A1+Dn.*z.*A2; B=B1+Dn.*z.*B2; A2=A1; B2=B1; A1=A; B1=B;
end
ft=ft+A./B; ft=sum(ft,3)-repmat(Fs(:,1,2),[1,M,1]);
ft=repmat(exp(c*t)/NT,[lv,1]).*ft; ft(:,1)=2*ft(:,1);
switch depict
    case 'p1', plott1(t,ft); case 'p2', plott2(t,ft);
    case 'p3', plott3(t,ft); otherwise display('Invalid Plot');
end
```

Table 5. Matlab listing of vector version of 1D NILT method

Here one more parameter depict is used to define a method of plotting individual items from a set of originals. The 1D NILT function is called as niltcv('F',tm,'depict'); where 'depict' is a text string 'p1', 'p2', or 'p3', see Tab. 6 for more details.

```
% --- Plotting functions called by 1D NILT, vector version ----
%---------- Multiple plotting into single figure -------------
function plott1(t,ft)
figure; plot(t,real(ft)); grid on;
figure; plot(t,imag(ft)); grid on;                    % optional
% ------------ Plotting into separate figures ----------------
function plott2(t,ft)
for k=1:size(ft,1)
    figure; plot(t,real(ft(k,:))); grid on;
    figure; plot(t,imag(ft(k,:))); grid on;           % optional
end
% ------------------- Plotting into 3D graphs ----------------
function plott3(t,ft)
global x;                                % x must be global in F
    m=length(t); tgr=[1:m/64:m,m];       % 65 time points chosen
    figure; mesh(t(tgr),x,real(ft(:,tgr)));
    figure; mesh(t(tgr),x,imag(ft(:,tgr)));           % optional
```

Table 6. Matlab listing of plotting functions for vector version of 1D NILT method

To get e.g. the right part of Fig. 8, that is the voltage responses of the circuit in Fig. 7, the calling the 1D NILT function looks like `niltcv('V4',1,'p1');` where `V4` denotes a name of the function defining individual responses as follows:

```
function f=V4(s)
I0=1e-3; C=1e-3; G=2e-2;
f(1,:)=I0./s./(G+s*C);
f(2,:)=I0./(s+5)./(G+s*C);
f(3,:)=2*pi*I0./(s.^2+4*pi^2)./(G+s*C);
f(4,:)=s.*I0./(s.^2+4*pi^2)./(G+s*C);
```

In this case the lines causing the imaginary parts plotting can be inactivated. The remaining plotting functions will be explained in the next chapter.

3.1.2 Application for transmission line simulation
Here, the 1D NILT algorithms will be used to simulate voltage and/or current distributions along transmission lines (TL), as shown on a Laplace-domain TL model in Fig. 9. As is well known, this model results from the application of a Laplace transform, with respect to time, on a pair of partial differential equations (telegraphic) of the form

$$-\frac{\partial v(t,x)}{\partial x} = R_0 i(t,x) + L_0 \frac{\partial i(t,x)}{\partial t} \ , \quad -\frac{\partial i(t,x)}{\partial x} = G_0 v(t,x) + C_0 \frac{\partial v(t,x)}{\partial t} \ , \tag{33}$$

with R_0, L_0, G_0, and C_0 as per-unit-length (p.-u.-l.) parameters, being constant for uniform TLs, and with a length l.

When considering zero initial voltage and current distributions, $v(0,x) = 0$ and $i(0,x) = 0$, and incorporating boundary conditions, we get the Laplace-domain solution in the forms

$$V(s,x) = V_i(s) \frac{Z_c(s)}{Z_i(s) + Z_c(s)} \cdot \frac{e^{-\gamma(s)x} + \rho_2(s)e^{-\gamma(s)[2l-x]}}{1 - \rho_1(s)\rho_2(s)e^{-2\gamma(s)l}} \ , \tag{34}$$

$$I(s,x) = V_i(s) \frac{1}{Z_i(s) + Z_c(s)} \cdot \frac{e^{-\gamma(s)x} - \rho_2(s)e^{-\gamma(s)[2l-x]}}{1 - \rho_1(s)\rho_2(s)e^{-2\gamma(s)l}} \ , \tag{35}$$

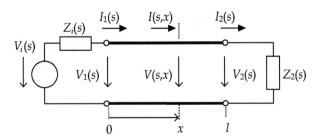

Fig. 9. Laplace-domain model of transmission line with linear terminations

where $Z_c(s)$ and $\gamma(s)$ are a characteristic impedance and a propagation constant, respectively,

$$Z_c(s) = \sqrt{\frac{R_0 + sL_0}{G_0 + sC_0}} \quad , \quad \gamma(s) = \sqrt{(R_0 + sL_0)(G_0 + sC_0)} \quad , \tag{36}$$

and $\rho_1(s)$ and $\rho_2(s)$ are reflection coefficients at the TL beginning and end, respectively,

$$\rho_1(s) = \frac{Z_i(s) - Z_c(s)}{Z_i(s) + Z_c(s)} \quad , \quad \rho_2(s) = \frac{Z_2(s) - Z_c(s)}{Z_2(s) + Z_c(s)} \quad . \tag{37}$$

In a general case of lossy TLs, the time-domain solutions cannot be found by an analytical method, thus the only way is to use some numerical technique.

As an example, let us consider the TL of a length $l = 1m$, with p.-u.-l. parameters $R_0 = 1m\Omega$, $L_0 = 600nH$, $G_0 = 2mS$, and $C_0 = 80pF$, terminated by resistive elements $Z_i = 10\Omega$, $Z_2 = 1k\Omega$, and excited by the voltage source waveform $v_i(t) = \sin^2(\pi t / 2 \cdot 10^{-9})$, $0 \le t \le 2 \cdot 10^{-9}$, and $v_i(t) = 0$, otherwise, with the Laplace transform

$$V_i(s) = \frac{2\pi^2 \left[1 - \exp(-2 \cdot 10^{-9} s) \right]}{s \left[(2 \cdot 10^{-9} s)^2 + 4\pi^2 \right]} \quad . \tag{38}$$

The Fig. 10 shows time dependences at the beginning, the centre, and the end of the TL, while the 1D NILT is called as `niltcv('Vs',4e-8,'p1');` where the function Vs is defined as

```
function f=Vs(s)
l=1; x=[0,1/2,1];
Ro=1e-3; Lo=600e-9; Go=2e-3; Co=80e-12;
Zi=10; Z2=1e3;
Vi=2*pi^2*(1-exp(-2e-9*s))./s./((2e-9*s).^2+4*pi^2);
Z=Ro+s*Lo; Y=Go+s*Co; Zc=sqrt(Z./Y); gam=sqrt(Z.*Y);
ro1=(Zi-Zc)./(Zi+Zc); ro2=(Z2-Zc)./(Z2+Zc);
Ks=Vi./(Zi+Zc)./(1-ro1.*ro2.*exp(-2*gam*l));
for k=1:length(x)
    f(k,:)=Ks.*Zc.*(exp(-gam*x(k))+ro2.*exp(-gam*(2*l-x(k))));
end
```

Similarly, current waveforms can be computed by the above function slightly modified according to (35). Both waveforms are depicted in Fig. 10.

Finally, it will be shown, how to obtain three-dimensional graphical results representing voltage and current distributions along the TL. Besides a possibility to use the `for` cycle, as shown in the function Vs above, another method based on 3D arrays will be applied, see the function Vsx below:

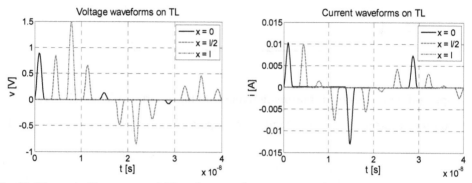

Fig. 10. Numerically computed TL voltage and current waveforms

```
function f=Vsx(s)
global x;
l=1;
Ro=1e-3; Lo=600e-9; Go=2e-3; Co=80e-12;
Zi=10; Z2=1e3;
x=linspace(0,1,65);        % 65 points along TL chosen
[S,X]=meshgrid(s,x);
Vi=2*pi^2*(1-exp(-2e-9*S))./S./((2e-9*S).^2+4*pi^2);
Z=Ro+S*Lo; Y=Go+S*Co; Zc=sqrt(Z./Y); gam=sqrt(Z.*Y);
ro1=(Zi-Zc)./(Zi+Zc); ro2=(Z2-Zc)./(Z2+Zc);
Ks=Vi./(Zi+Zc)./(1-ro1.*ro2.*exp(-2*gam*l));
f=Ks.*Zc.*(exp(-gam.*X)+ro2.*exp(-gam.*(2*l-X)));
```

In this case, the 1D NILT algorithm in Tab. 5 is called as niltcv('Vsx',2e-8,'p3'); that is the plott3 function is used for the plotting, see Tab. 6, and a time limit is half of that in Fig. 10 to get well-observable results. Again, the current distributions can be gained via the above function slightly modified according to (35). Both 3D graphs are depicted in Fig. 11.

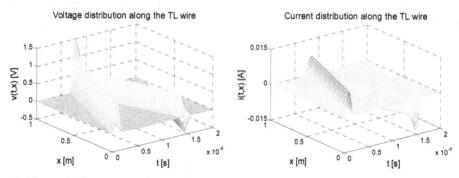

Fig. 11. Numerically computed TL voltage and current distributions

3.2 Two-dimensional NILT algorithm application

A two-dimensional Laplace transform can generally be used for the solution of linear partial differencial equations with two variables. The advantage is that we get completely algebraic equations leading to much easier solution in the Laplace domain. A final step in the solution is then the utilization of the 2D NILT algorithm to get results in the original domain. Such a possibility will be shown on the example of telegraphic equations describing transmission lines, and results will be compared with the 1D NILT approach.

3.2.1 Application for transmission line simulation

Herein, rather less conventional approach for the simulation of voltage and/or current distributions along the TLs will be discussed. As is obvious from the telegraphic equations (33) they can be transformed not only with respect to the time t, which was matter of the previous paragraph, but also with respect to the geometrical coordinate x to get completely algebraic equations. After performing such the Laplace transforms, incorporating boundary conditions given by the terminating circuits, and considering again zero initial voltage and current distributions, $v(0,x) = 0$ and $i(0,x) = 0$, we get (Valsa & Brančík, 1998b)

$$V(s,q) = \frac{qV_1(s) - \gamma(s)Z_c(s)I_1(s)}{q^2 - \gamma^2(s)}, \tag{39}$$

$$I(s,q) = \frac{qI_1(s) - \dfrac{\gamma(s)}{Z_c(s)}V_1(s)}{q^2 - \gamma^2(s)}, \tag{40}$$

where $V_1(s) = V(s,0)$ and $I_1(s) = I(s,0)$ are given by (34) and (35), respectively, see also Fig. 9. Thus the 2D NILT function according to Tab. 3 can be called as nilt2c('Vsq',2e-8,1); leading to the same graphical results as are shown in Fig. 11. The function Vsq can be of the form as stated below. The current distribution is obtained via the same function slightly modified according to (40).

```
function f=Vsq(s,q)
l=1; Zi=10; Z2=1e3;
Ro=1e-3; Lo=600e-9; Go=2e-3; Co=80e-12;
Vi=2*pi^2*(1-exp(-2e-9*s))./s./((2e-9*s).^2+4*pi^2);
Z=Ro+s*Lo; Y=Go+s*Co; Zc=sqrt(Z./Y); gam=sqrt(Z.*Y);
ro1=(Zi-Zc)./(Zi+Zc); ro2=(Z2-Zc)./(Z2+Zc);
Ks=Vi./(Zi+Zc)./(1-ro1.*ro2.*exp(-2*gam*l));
V1=Ks.*Zc.*(1+ro2.*exp(-2*gam*l));
I1=Ks.*(1-ro2.*exp(-2*gam*l));
f=(q.*V1-Zc.*gam.*I1)./(q.^2-gam.^2);
```

One can notice an interesting thing, namely getting both voltage and current graphs by a single computation step. It is enabled by putting together the voltage and current transforms forming respectively real and imaginary parts of an artificial complex transform, and letting active the program command for the plotting the imaginary part of the original function, see Tab. 3. In our example, if the bottom line in the Vsq function is changed to

```
f=((q.*V1-Zc.*gam.*I1)+j*(q.*I1-gam./Zc.*V1))./(q.^2-gam.^2);
```

then both graphs in Fig. 11 are obtained simultaneously. The same possibility exists for the 1D NILT functions discussed earlier. There is no obvious physical meaning of such articifial complex transfoms, it is only a formal tool for inverting two transforms in parallel instead.

3.3 Multidimensional LT in nonlinear electrical circuits simulation

As is known some classes of nonlinear systems can be described through a Volterra series expansion, accurately enough from the practical point of view, when a response $v(t)$ to a stimulus $i(t)$ has a form (Schetzen, 2006)

$$v(t) = \sum_{n=1}^{\infty} v_n(t), \qquad (41)$$

where the terms of the infinite sum are

$$v_n(t) = \underbrace{\int_{-\infty}^{\infty} \cdots \int_{-\infty}^{\infty}}_{n-fold} h_n(\tau_1, \tau_2, \ldots, \tau_n) \prod_{k=1}^{n} i(t - \tau_k) d\tau_k, \qquad (42)$$

with $h_n(\tau_1, \tau_2, \ldots, \tau_n)$ as an n-th order Volterra kernel, called as a nonlinear impulse response as well. The Fig. 12 shows these equations in their graphical form.

By introducing new variables, $t_1 = t_2 = \ldots = t_n = t$, and by using the n-dimensional Laplace transform (1), the n-fold convolution integral (42) leads to a Laplace domain response

$$V_n(s_1, s_2, \cdots, s_n) = H_n(s_1, s_2, \cdots, s_n) \prod_{k=1}^{n} I(s_k), \qquad (43)$$

with $H_n(s_1, s_2, \ldots, s_n)$ as a nonlinear transfer function.

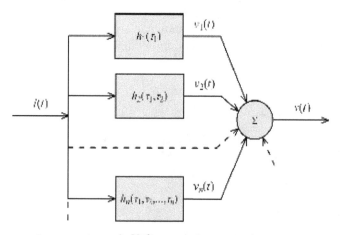

Fig. 12. Nonlinear system response via Volterra series expansion

A few methods are at disposal to find the transfer function for a given nonlinear system, like a harmonic input method, e.g. (Bussgang at al., 1974; Karmakar, 1980). Further procedure is usually based on the association of variables, (J. Debnath & N.C. Debnath, 1991; Reddy & Jagan, 1971), transforming $V_n(s_1, s_2, \ldots, s_n)$ into the function of a single variable $V_n(s)$, and enabling to use a one-dimensional ILT to get the required terms $v_n(t)$ in (41). In contrast to this procedure, it is also possible to determine the terms $v_n(t_1, t_2, \ldots, t_n)$ by the use of the n-dimensional ILT, considering $t_1 = t_2 = \ldots = t_n = t$ in the result as a final step. That is why the above NILT procedures can be adapted in this respect being able to serve as a tool for the nonlinear circuits transient simulation.

3.3.1 Utilization of 1D to 3D NILTs for nearly nonlinear circuits

The utilization of the NILT methods developed, up to three-dimensional case, will be shown on the solution of a nearly nonlinear circuit in Fig. 13. As can be observed this is just Fig. 7 modified to introduce a nonlinearity via G_2 conductance.

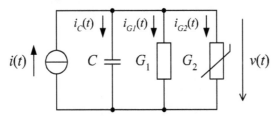

Fig. 13. Electrical circuit with nonlinear resistive element G_2

Assuming a square nonlinearity, a circuit equation is

$$C\frac{dv(t)}{dt} + G_1 v(t) + G_2 v^2(t) = i(t) \cdot \tag{44}$$

By using the harmonic input method, and limiting the solution on the first three terms only, we get the nonlinear transfer functions for (43) as

$$H_1(s_1) = (s_1 C + G_1)^{-1}, \tag{45}$$

$$H_2(s_1, s_2) = -G_2 H_1(s_1) H_1(s_2) H_1(s_1 + s_2), \tag{46}$$

$$H_3(s_1, s_2, s_3) = -\frac{G_2}{3}[H_1(s_1)H_2(s_2, s_3) + H_1(s_2)H_2(s_1, s_3) + H_1(s_3)H_2(s_1, s_2)]H_1(s_1 + s_2 + s_3). \tag{47}$$

Let us use an exciting current and its Laplace transform as

$$i(t) = I_0 e^{-at} 1(t) \quad \mapsto \quad I(s) = \frac{I_0}{s+a}, \tag{48}$$

$a \geq 0$. The substitution (45) – (48) into (43) gives us respective Laplace-domain responses which will undergo the 1D, 2D and 3D NILT algorithms, respectively. We can write

$$v(t) = v_1(t) + v_2(t) + v_3(t) = v_1(t_1)\big|_{t_1 = t} + v_2(t_1, t_2)\big|_{t_1 = t_2 = t} + v_3(t_1, t_2, t_3)\big|_{t_1 = t_2 = t_3 = t} =$$
$$= \mathbb{L}_1^{-1}[V_1(s_1)]\big|_{t_1 = t} + \mathbb{L}_2^{-1}[V_2(s_1, s_2)]\big|_{t_1 = t_2 = t} + \mathbb{L}_3^{-1}[V_3(s_1, s_2, s_3)]\big|_{t_1 = t_2 = t_3 = t} \tag{49}$$

with $\mathbb{L}_k^{-1}[.]$ as a k-dimensional ILT. Thereby, a time-consuming association of variables can be omitted, e.g. (Wambacq & Sansen, 2010). Individual terms $v_k(t)$ are depicted in Fig. 14, for values agreeing with the linear circuit version in Fig. 7. The current $i(t)$ is defined by $a = 0$ (a unit step), and $a = 5$ (an exponential impuls), compare the first two columns in Tab. 4.

The resultant voltage responses computed according to (49) are shown in Fig. 15, including relative errors, where also dependences on Volterra series orders are presented.

The relative errors above were computed via a Matlab ODE45 Runge-Kutta function applied directly to the nonlinear ODE (44). As expected, more Volterra terms lead to more accurate results, see also (Brančík, 2009) where up to 2nd-order terms were considered, and respective Matlab listings are presented.

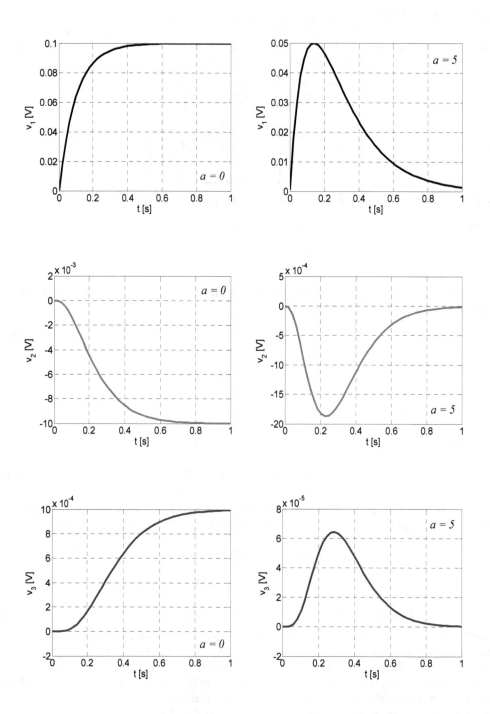

Fig. 14. Numerical inversions leading to voltage response Volterra series terms

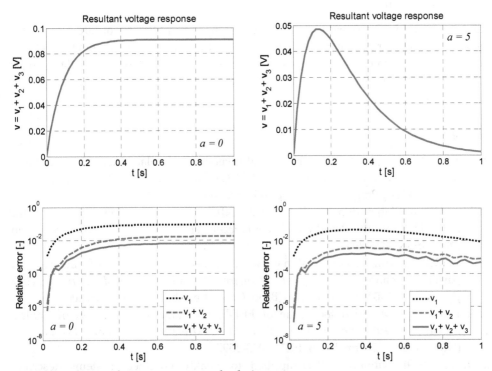

Fig. 15. Resultant voltage responses and relative errors

4. Conclusion

The paper dealt with a specific class of techniques for the numerical inversion of Laplace transforms, namely based on a complex Fourier series approximation, and connected with a quotient-difference algorithm to accelerate the convergence of infinite series arising in the approximate formulae. The 1D to 3D NILT techniques have been programmed in the Matlab language (R2007b), and most important ones provided as the Matlab function listings. To guide readers all the algorithms were explained on selected examples from field of electrical engineering, including right callings of the functions. In contrast to most others the NILT methods here developed are utilizable to numerically invert complex Laplace transforms, leading to complex originals, which can be useful for some special purposes. As has resulted from error analyses the accuracies range relative errors from 10^{-8} to 10^{-10} without difficulties which is acceptable for most of practical needs. Based on Matlab functions presented, one could further generalize a vector version of the 1D NILT function towards a matrix one, enabling e.g. to simulate multiconductor transmission line systems, as is shown in (Brančík, 1999), where, however, an alternative technique, so-called ε algorithm, has been applied to accelerate the convergence of infinite series. According to the author's knowledge, the paper presented ranks among few summary works describing multidimensional NILT techniques, covering Matlab listings beyond, based just on a complex Fourier series approximation, and in conjunction with the quotient-difference algorithm, which seems to be more numerically stable compared to the ε algorithm mentioned above.

5. Acknowledgment

Research described in this paper was supported by the Czech Ministry of Education under the MSM 0021630503 research project MIKROSYN, the European Community's Seventh Framework Programme under grant agreement no. 230126, and the project CZ.1.07/2.3.00/20.0007 WICOMT of the operational program Education for competitiveness.

6. References

Brančík, L. (1999). Programs for fast numerical inversion of Laplace transforms in Matlab language environment. *Proceedings of 7th Conference MATLAB'99*, pp. 27-39, ISBN 80-7080-354-1, Prague, Czech Republic, November 3, 1999

Brančík, L. (2005). Elaboration of FFT-based 2D-NILT methods in terms of accuracy and numerical stability. *Przeglad Elektrotechniczny*, Vol. 81, No. 2, (February 2005), pp. 84-89, ISSN 0033-2097

Brančík, L. (2007a). Numerical Inversion of two-dimensional Laplace transforms based on partial inversions. *Proceedings of 17th International Conference Radioelektronika 2007*, pp. 451-454, ISBN 1-4244-0821-0, Brno, Czech Republic, April 24-25, 2007

Brančík, L. (2007b). Modified technique of FFT-based numerical inversion of Laplace transforms with applications. *Przegląd Elektrotechniczny*, Vol. 83, No. 11, (November 2007), pp. 53-56, ISSN 0033-2097

Brančík, L. (2009). Numerical ILTs applied to weakly nonlinear systems described by second-order Volterra series. *ElectroScope*, [online], Special Issue on Conference EDS 2009, 4 pages, Available from http://electroscope.zcu.cz, ISSN 1802-4564

Brančík, L. (2010a). Utilization of NILTs in simulation of nonlinear systems described by Volterra series. *Przeglad Elektrotechniczny*, Vol. 86, No. 1, (January 2010), pp. 68-70, ISSN 0033-2097

Brančík, L. (2010b). Numerical inversion of 3D Laplace transforms for weakly nonlinear systems solution. *Proceedings of 20th International Conference Radioelektronika 2010*, pp. 221-224, ISBN 978-1-4244-6319-0, Brno, Czech Republic, April 19-21, 2010

Brančík, L. (2010c). Technique of 3D NILT based on complex Fourier series and quotient-difference algorithm. *Proceedings of 2010 IEEE International Conference on Electronics, Circuits, and Systems ICECS2010*, pp. 207-210, ISBN 978-1-4244-8156-9, Athens, Greece, December 12-15, 2010

Brančík, L. (2011). Error analysis at numerical inversion of multidimensional Laplace transforms based on complex Fourier series approximation. *IEICE Transactions on Fundamentals of Electronics, Communications and Computer Sciences*, Vol. E94-A, No. 3, (March 2011), p. 999-1001, ISSN 0916- 8508

Bussgang, J.J.; Ehrman, L. & Graham, J.W. (1974). Analysis of nonlinear systems with multiple inputs. *Proceedings of the IEEE*, Vol. 62, No. 8, (August 1974), pp. 1088-1119, ISSN 0018-9219

Cohen, A.M. (2007). *Numerical methods for Laplace transform inversion*. Springer Science, ISBN 978-0-387-28261-9, New York, U.S.A

Debnath, J. & Debnath, N.C. (1991). Associated transforms for solution of nonlinear equations. *International Journal of Mathematics and Mathematical Sciences,* Vol. 14, No. 1, (January 1991), pp. 177-190, ISSN 0161-1712

DeHoog, F.R.; Knight, J.H. & Stokes, A.N. (1982). An improved method for numerical inversion of Laplace transforms. *SIAM Journal on Scientific and Statistical Computing,* Vol. 3, No. 3, (September 1982), pp. 357-366, ISSN 0196-5204

Hwang, C.; Guo, T.-Y. & Shih, Y.-P. (1983). Numerical inversion of multidimensional Laplace transforms via block-pulse functions. *IEE Proceedings D - Control Theory & Applications,* Vol. 130, No. 5, (September 1983), pp. 250-254, ISSN 0143-7054

Hwang, C. & Lu, M.-J. (1999). Numerical inversion of 2-D Laplace transforms by fast Hartley transform computations. *Journal of the Franklin Institute,* Vol. 336, No. 6, (August 1999), pp. 955-972, ISSN 0016-0032

Karmakar, S.B. (1980). Laplace transform solution of nonlinear differential equations. *Indian Journal of Pure & Applied Mathematics,* Vol. 11, No. 4, (April 1980), pp. 407-412, ISSN 0019-5588

Macdonald, J.R. (1964). Accelerated convergence, divergence, iteration, extrapolation, and curve fitting, *Journal of Applied Physics,* Vol. 35, No. 10, (February 1964), pp. 3034-3041, ISSN 0021-8979

McCabe, J.H. (1983). The quotient-difference algorithm and the Padé table: An alternative form and a general continued fraction. *Mathematics of Computation,* Vol. 41, No. 163, (July 1983), pp. 183-197, ISSN 0025-5718

Reddy, D.C. & Jagan, N.C. (1971). Multidimensional transforms: new technique for the association of variebles. *Electronics Letters,* Vol. 7, No. 10, (May 1971), pp. 278 – 279, ISSN 0013-5194

Rutishauser, H. (1957). *Der quotienten-differenzen-algorithmus.* Birkhäuser Verlag, Basel, Schweiz

Schetzen, M. (2006). *The Volterra and Wiener theories of nonlinear systems.* Krieger Publishing, ISBN 978-1-575-24283-5, Melbourne, Florida, U.S.A

Singhal, K.; Vlach, J. & Vlach, M. (1975). Numerical inversion of multidimensional Laplace transform. *Proceedings of the IEEE,* Vol. 63, No. 11, (November 1975), pp. 1627-1628, ISSN 0018-9219

Valsa, J. & Brančík, L. (1998a). Approximate formulae for numerical inversion of Laplace transforms. *International Journal of Numerical Modelling: Electronic Networks, Devices and Fields,* Vol. 11, No. 3, (May-June 1998), pp. 153-166, ISSN 0894-3370

Valsa, J. & Brančík, L. (1998b). Time-domain simulation of lossy transmission lines with arbitrary initial conditions. *Proceedings of Advances in Systems, Signals, Control and Computers,* Vol. III, pp. 305-307, ISBN 0-620-23136-X, Durban, South Africa, September 22-24, 1998

Wambacq, P. & Sansen, W.M.C. (2010). *Distortion analysis of analog integrated circuits.* Kluwer Academic Publishers, ISBN 978-1-4419-5044-4, Boston, U.S.A

Wu, J.L.; Chen, C.H. & Chen, C.F. (2001). Numerical inversion of Laplace transform using Haar wavelet operational matrices. *IEEE Transactions on Circuits and Systems – I: Fundamental Theory and Applications,* Vol. 48, No. 1, (January 2001), pp. 120-122, ISSN 1057-7122

2

Control of Distributed Parameter Systems- Engineering Methods and Software Support in the MATLAB & Simulink Programming Environment

Gabriel Hulkó, Cyril Belavý, Gergely Takács,
Pavol Buček and Peter Zajíček
Institute of Automation, Measurement and Applied
Informatics Faculty of Mechanical Engineering
Center for Control of Distributed Parameter Systems
Slovak University of Technology Bratislava
Slovak Republic

1. Introduction

Distributed parameter systems (DPS) are systems with state/output quantities $X(\mathbf{x},t)$ /$Y(\mathbf{x},t)$ – parameters which are defined as quantity fields or infinite dimensional quantities distributed through geometric space, where $\mathbf{x}$ – in general is a vector of the three dimensional Euclidean space. Thanks to the development of information technology and numerical methods, engineering practice is lately modelling a wide range of phenomena and processes in virtual software environments for numerical dynamical analysis purposes such as ANSYS - www.ansys.com, FLUENT (ANSYS Polyflow) - www.fluent.com , ProCAST www.esi-group.com/products/casting/, COMPUPLAST - www.compuplast.com, SYSWELD – www.esi-group.com/products/welding, COMSOL Multiphysics - www.comsol.com, MODFLOW, MODPATH,... www.modflow.com , STAR-CD - www.cd-adapco.com, MOLDFLOW - www.moldflow.com, ... Based on the numerical solution of the underlying partial differential equations (PDE) these virtual software environments offer colorful, animated results in 3D. Numerical dynamic analysis problems are solved both for technical and non-technical disciplines given by numerical models defined in complex 3D shapes. From the viewpoint of systems and control theory these dynamical models represent DPS. A new challenge emerges for the control engineering practice, which is the objective to formulate control problems for dynamical systems defined as DPS through numerical structures over complex spatial structures in 3D.

The main emphasis of this chapter is to present a philosophy of the engineering approach for the control of DPS - given by numerical structures, which opens a wide space for novel applications of the toolboxes and blocksets of the MATLAB & Simulink software environment presented here.

The first monographs in the field of DPS control have been published in the second half of the last century, where mathematical foundations of DPS control have been established. This mathematical theory is based on analytical solutions of the underlying PDE (Butkovskij, 1965; Lions, 1971; Wang, 1964). That is the decomposition of dynamics into time and space components based on the eigenfunctions of the PDE. Recently in the field of mathematical control theory of DPS, publications on control of PDE have appeared (Lasiecka & Triggiani, 2000;...).

An engineering approach for the control of DPS is being developed since the eighties of the last century (Hulkó et al., 1981-2010). In the field of lumped parameters system (LPS) control, where the state/output quantities $x(t)/y(t)$ – parameters are given as finite dimensional vectors, the actuator together with the controlled plant make up a controlled LPS. In this sense the actuators and the controlled plant as a DPS create a controlled lumped-input and distributed-parameter-output system (LDS). In this chapter the general decomposition of dynamics of controlled LDS into time and space components is introduced, which is based on numerically computed distributed parameter transient and impulse characteristics given on complex shape definition domains in 3D. Based on this decomposition a methodical framework of control synthesis decomposition into space and time tasks will be presented where in space domain approximation problems are solved and in time domain synthesis of control is realized by lumped parameter control loops. For the software support of modelling, control and design of DPS, the **Distributed Parameter Systems Blockset for MATLAB & Simulink (DPS Blockset)** - www.mathworks.com/products/connections/ has been developed within the CONNECTIONS program framework of The MathWorks, as a Third-Party Product of The MathWorks Company (Hulkó et al., 2003-2010). When solving problems in the time domain, toolboxes and blocksets of the MATLAB & Simulink software environment such as for example the Control Systems Toolbox, Simulink Control Design, System Identification Toolbox, etc. are utilized. In the space relation the approximation task is formulated as an optimization problem, where the Optimization Toolbox is made use of. A web portal named **Distributed Parameter Systems Control** - www.dpscontrol.sk has been created for those interested in solving problems of DPS control (Hulkó et al., 2003-2010). This web portal features application examples from different areas of engineering practice such as the control of technological and manufacturing processes, mechatronic structures, groundwater remediation etc. Moreover this web portal offers the demo version of the **DPS Blockset** with the **Tutorial, Show, Demos** and **DPS Wizard** for download, along with the **Interactive Control** service for the interactive solution of model control problems via the Internet.

2. DPS – DDS – LDS

Generally in the control of lumped parameter systems the actuator and the controlled plant create a lumped parameter controlled system. In the field of DPS control the actuators together with the controlled plant - generally being a distributed-input and distributed-parameter-output system (DDS) create a controlled lumped-input and distributed-parameter-output system (LDS). Fig. 1.-3. and Fig. 6., 11., 14.

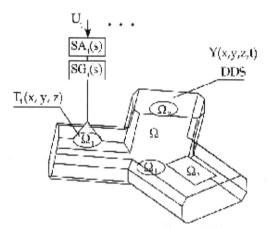

Fig. 1. Controlled DPS as LDS - heating of metal body of complex-shape

$\{SA_i(s)\}_i$, $\{SG_i(s)\}_i$, $\{T_i(x,y,z)\}_i$ - models of actuators

DDS   - distributed-input and distributed-parameter-output system

$\{U_i\}_i$ - lumped actuating quantities

Ω / $\{\Omega_i\}_i$ - complex-shape definition domain in 3D / actuation subdomains

$Y(x,y,z,t)$ – temperature field – distributed output quantity

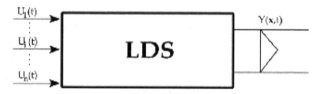

Fig. 2. Lumped-input and distributed-parameter-output system – LDS

$\{U_i(t)\}_i$ – lumped input quantities

$Y(x,t) = Y(x,y,z,t)$ – distributed output quantity

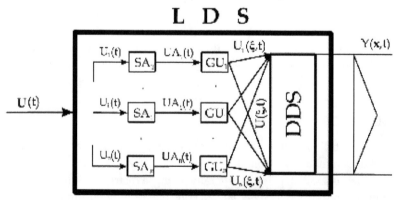

Fig. 3. General structure of lumped-input and distributed-parameter-output systems

LDS - lumped-input and distributed-parameter-output system

$\{SA_i\}_i$ - actuating members of lumped input quantities

$\{GU_i\}_i$ - generators of distributed input quantities

DDS - distributed-input and distributed-parameter-output system

$\mathbf{U}(t) = \{U_i(t)\}_i$ - vector of lumped input quantities of LDS

$\{UA_i(t)\}_i$ - output quantities of lumped parameter actuators

$\{U_i(\xi,t)\}_i$ - distributed output quantities of generators $\{GU_i\}_i$

$U(\xi,t)$ - overall distributed input quantity for DDS

$Y(\mathbf{x},t) = Y(x,y,z,t)$ - distributed output quantity

Input-output dynamics of these DPS can be described, from zero initial conditions, by

$$Y(\mathbf{x},t) = \sum_{i=1}^{n} Y_i(\mathbf{x},t) = \sum_{i=1}^{n} \mathcal{G}_i(\mathbf{x},t) \otimes U_i(t) = \sum_{i=1}^{n} \int_0^t \mathcal{G}_i(\mathbf{x},\tau) U_i(t-\tau) d\tau \qquad (1)$$

or in discrete form

$$Y(\mathbf{x},k) = \sum_{i=1}^{n} Y_i(\mathbf{x},k) = \sum_{i=1}^{n} \mathcal{G}H_i(\mathbf{x},k) \oplus U_i(k) = \sum_{i=1}^{n} \sum_{q=0}^{k} \mathcal{G}H_i(\mathbf{x},q) U_i(k-q) \qquad (2)$$

where $\otimes$ marks convolution product and $\oplus$ marks convolution sum, $\mathcal{G}_i(\mathbf{x},t)$ – distributed parameter impulse response of LDS to the i-th input, $\mathcal{G}H_i(\mathbf{x},k)$ – discrete time (DT) distributed parameter impulse response of LDS with zero-order hold units H - HLDS to the i-th input, $Y_i(\mathbf{x},t)$ - distributed output quantity of LDS to the i-th input, $Y_i(\mathbf{x},k)$ – DT distributed output quantity of HLDS to the i-th input. For simplicity in this chapter distributed quantities are considered mostly as continuous scalar quantity fields with unit sampling interval in time domain. Whereas DT distributed parameter step responses $\{\mathcal{H}H_i(\mathbf{x},k)\}_{i,k}$ of HLDS can be computed by common analytical or numerical methods then DT distributed parameter impulse responses can be obtained as

$$\{\mathcal{G}H_i(\mathbf{x},k) = \mathcal{H}H_i(\mathbf{x},k) - \mathcal{H}H_i(\mathbf{x},k\text{-}1)\}_{i,k} \qquad (3)$$

3. Decomposition of dynamics

The process of dynamics decomposition shall be started from DT distributed parameter step and impulse responses of the analysed LDS. For an illustration, procedure of decomposition of dynamics and control synthesis will be shown on the LDS with zero-order hold units H – HLDS – distributed only on the interval $[0,L]$, with output quantity

$$Y(x,k) = \sum_{i=1}^{n} Y_i(x,k)$$ discretised in time relation and continuous in space relation on this

interval. Nevertheless the following results are valid in general both for continous or discrete distributed quantities in space relation given on compex-shape definition domains over 3D as well.

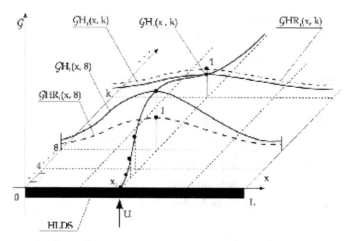

Fig. 4. i-th discrete distributed parameter impulse response of HLDS

$\mathcal{G}H_i(x_i,k)$ - partial DT impulse response in time, t - relation to the i-th input considered as response with maximal amplitude in point $x_i \in [0,L]$

$\{\mathcal{G}H_i(x,k)\}_{i,k}$ - partial DT impulse responses to the i-th input in space - x - relation

$\{\mathcal{G}HR_i(x,k)\}_{i,k}$ – reduced partial DT impulse responses to the i-th input in space,

x – relation for timesteps $\{k\}_k$

If the reduced DT partial distributed parameter impulse responses are defined as

$$\left\{ \mathcal{G}HR_i(x,k) = \frac{\mathcal{G}H_i(x,k)}{\mathcal{G}H_i(x_i,k)} \right\}_{i,k} \tag{4}$$

for $\{\mathcal{G}H_i(x_i,k) \neq 0\}_{i,k}$, then the i-th DT distributed output quantiy in (2) can be rewritten by the means of the reduced characteristics as follows

$$Y_i(x,k) = \mathcal{G}H_i(x_i,k)\mathcal{G}HR_i(x,k) \oplus U_i(k) \tag{5}$$

At fixed x_i the partial DT distributed output quantity in time direction $Y_i(x_i,k)$ is given as

the convolution sum $\mathcal{G}H_i(x_i,k) \oplus U_i(k) = \sum_{q=0}^{k} \mathcal{G}H_i(x_i,q)U_i(k-q)$, in case the relation

$\{\mathcal{G}HR_i(x_i,q) = 1\}_{q=0,k}$ holds at the fixed point x_i. At fixed k, the partial discrete distributed

output quantity in space direction $Y_i(x,k)$ is given as a linear combination of elements

$\{\mathcal{G}HR_i(x,q)\}_{q=0,k}$, where the reduced discrete partial distributed characteristics

$\{\mathcal{G}HR_i(x,q)\}_{q=0,k}$ are multiplied by corresponding elements of the set

$\{\mathcal{G}H_i(x_i,q)U_i(k-q)\}_{q=0,k}$., see Fig. 5.

This decomposition is valid for all given lumped input $\{U_i\}_i$ and corresponding output quantities $\{Y_i(x,k)\}_{i,k}$ - thus we obtain time and space components of HLDS dynamics in the following form:

Time Components of Dynamics $\{GH_i(x_i,k)\}_{i,k}$ - for given i and chosen x_i - variable k

Space Components of Dynamics $\{GHR_i(x,k)\}_{i,k}$ - for given i and chosen k - variable x

Also reduced components of single distributed output quantities are

$$\{YR_i(x,k)\}_i = \left\{\frac{Y_i(x,k)}{Y_i(x_i,k)}\right\}_{i,k} \tag{6}$$

then $\{Y_i(x_i,k) \neq 0\}_{i,k}$ can be considered as time components and $\{YR_i(x,k)\}_{i,k}$ as space components of the output quantities.

When reduced steady-state distributed parameter transient responses are introduced $\{\mathcal{H}HR_i(x,\infty)\}_i = \{\mathcal{H}H_i(x,\infty)/\mathcal{H}H_i(x_i,\infty)\}_i$ - for $\{\mathcal{H}H_i(x_i,\infty) \neq 0\}_i$ - and discrete transfer functions $\{SH_i(x_i,z)\}_i$ are assigned to partial distributed parameter transient responses with maximal amplitudes at points $\{x_i\}_i$ on the interval $[0,L]$, we obtain time and space components of HLDS dynamics for steady-state as:

Time Components of Dynamics $\{SH_i(x_i,z)\}_i$ - for given i and chosen x_i - variable z

Space Components of Dynamics $\{\mathcal{H}HR_i(x,\infty)\}_i$ - for given i in ∞ - variable x

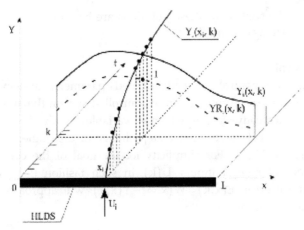

Fig. 5. Partial distributed output quantities in time and space direction

U_i - i-th DT lumped input quantity

$Y_i(x_i,k)$ - i-th partial DT distributed output quantity in time domain at chosen point x_i

$Y_i(x,k)/YR_i(x,k)$ - i-th partial distributed output/reduced output quantity in space direction

For the steady-state, when $k \to \infty$

$$\{YR_i(x,\infty) \to \mathcal{H}HR_i(x,\infty)\}_{i,k} \tag{7}$$

then in the steady-state

$$Y(x,\infty) = \sum_{i=1}^{n} Y_i(x_i,\infty)YR_i(x,\infty) = \sum_{i=1}^{n} Y_i(x_i,\infty)\mathcal{H}HR_i(x,\infty) \tag{8}$$

When distributed quantities are used in discrete form as finite sequences of quantities, the discretization in space domain is usually considered by the computational nodes of the numerical model of the controlled DPS over the compex-shape definition domain in 3D.

4. Distributed parameter systems of control

Based on decomposition of HLDS dynamics into time and space components, possibilities for control synthesis are also suggested by an analogous approach. In this section a methodical framework for the decomposition of control synthesis into space and time problems will be presented by select demonstration control problems. In the space domain control synthesis will be solved as a sequence of approximation tasks on the set of space components of controlled system dynamics, where distributed parameter quantities in particular sampling times are considered as continuous functions on the interval $[0,L]$ as elements of strictly convex normed linear space **X** with quadratic norm. It is necessary to note as above that the following results are valid in general for DPS given on compex-shape definition domains in 3D both for continous or discrete distributed quantities, in the space relation as well.

In the time domain, the control synthesis solutions are based on synthesis methods of DT lumped parameter systems of control.

4.1 Open-loop control

Assume the open-loop control of a distributed parameter system, where dynamic characteristics give an ideal representation of controlled system dynamics and $V(x,t) = 0$, that is with zero initial steady-state, in which all variables involved are equal to zero – see see Fig. 6 for reference. Let us consider a step change of distributed reference quantity - $W(x,k) = W(x,\infty)$, see Fig. 7. For simplicity let the goal of the control synthesis is to generate a sequence of control inputs $\bar{U}(k)$ in such fashion that in the steady-state, for $k \to \infty$, the control error $E(x,k) = W(x,\infty) - Y(x,k)$ will approach its minimal value $\breve{E}(x,\infty)$ in the quadratic norm:

$$\min\|E(x,\infty)\| = \min\|W(x,\infty) - Y(x,\infty)\| = \|\breve{E}(x,\infty)\| \tag{9}$$

First, an approximation problem will be solved in the space synthesis (SS) block:

$$\min\left\|W(x,\infty) - \sum_{i=1}^{n} W_i(x_i,\infty)\mathcal{H}HR_i(x,\infty)\right\| \tag{10}$$

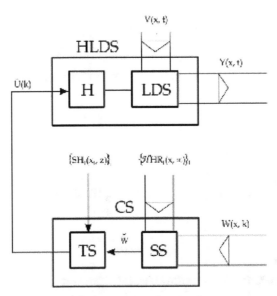

Fig. 6. Distributed parameter open-loop system of control

LDS   - lumped-input and distributed-parameter-output system
H   - zero-order hold units
HLDS   - controlled system with zero-order hold units
CS   - control synthesis
TS   - time part of control synthesis
SS   - space part of control synthesis

$Y(x,t) / W(x,k) = W(x,\infty)$ - distributed controlled/reference quantity

$V(x,t)$ - distributed disturbance quantity

$\breve{W} = \left\{ \breve{W}_i(x_i,\infty) \right\}_i$ - vector of lumped reference quantities

$\bar{U}(k)$ - vector of lumped control quantities

$\left\{ SH_i(x_i,z) / \mathcal{H}HR_i(x,\infty) \right\}_i$ - time/space components of controlled system dynamics

where $\left\{ \mathcal{H}HR_i(x,\infty) \right\}_i$ are steady-state reduced distributed parameter transient responses of the controlled system – HLDS and $\left\{ W_i(x_i,\infty) \right\}_i$ are parameters of approximation. Functions $\left\{ \mathcal{H}HR_i(x,\infty) \right\}_i$ form a finite-dimensional subspace of approximation functions *Fn* in the strictly convex normed linear space of distributed parameter quantities **X** on $[0,L]$ with quadratic norm, where the approximation problem is to be solved, see Fig. 8. for reference. From approximation theory in this relation is known the theorem:
*Let Fn be a finite-dimensional subspace of a strictly convex normed linear space **X**. Then, for each f∈ **X**, there exists a unique element of best approximation.*
(Shadrin, 2005). So the solution of the approximation problem (10) is guaranteed as a unique

best approximation $\breve{W}O(x,\infty) = \sum\limits_{i=1}^{n} \breve{W}_i(x_i,\infty)\mathcal{H}HR_i(x,\infty)$, where $\breve{\breve{W}} = \left\{ \breve{W}_i(x_i,\infty) \right\}_i$ is the

vector of optimal approximation parameters. Hence:

$$\min \left\| W(x,\infty) - \sum_{i=1}^{n} W_i(x_i,\infty) \mathcal{H}HR_i(x,\infty) \right\| =$$

$$= \left\| W(x,\infty) - \sum_{i=1}^{n} \breve{W}_i(x_i,\infty) \mathcal{H}HR_i(x,\infty) \right\| = \left\| W(x,\infty) - \breve{W}O(x,\infty) \right\|$$

(11)

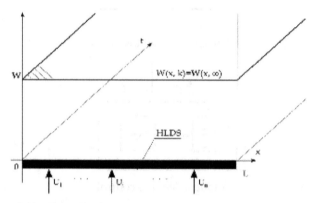

Fig. 7. Step change of distributed reference quantity

HLDS   - controlled system with zero-order hold units

$\{U_i\}_i$ - lumped control quantities

$W(x,k) = W(x,\infty)$ - step change of distributed reference quantity

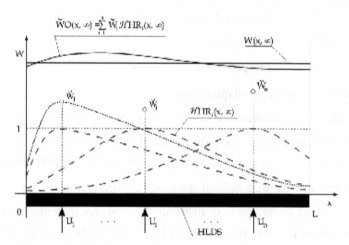

Fig. 8. Solution of the approximation problem

HLDS   - controlled system with zero-order hold units

$\{U_i\}_i$ - lumped control quantities

$\{\breve{W}_i\}_i = \{\breve{W}_i(x_i,\infty)\}_i$ - optimal approximation parameters, lumped references

$\{\mathcal{H}HR_i(x,\infty)\}_i$ - reduced steady-state distributed parameter transient responses

$W(x,\infty)$ - distributed reference quantity

$\breve{W}O(x,\infty)$ - unique best approximation of reference quantity

Let us assume vector $\breve{W}$ enters the block of time synthesis (TS). In this block, there are „n" single-input / single-output (SISO) lumped parameter control loops: $\left\{SH_i(x_i,z),R_i(z)\right\}_i$, see Fig. 9. for reference. The controlled systems of these loops are lumped parameter systems assigned to HLDS as time components of dynamics: $\left\{SH_i(x_i,z)\right\}_i$. Controllers, $\left\{R_i(z)\right\}_i$, are to be chosen such that for $k \to \infty$ the following relation holds:

$$\left\{\lim_{k\to\infty}\breve{E}_i(x_i,k)=\lim_{k\to\infty}\left[\breve{W}_i(x_i,\infty)-Y_i(x_i,k)\right]=0\right\}_{i,k} \tag{12}$$

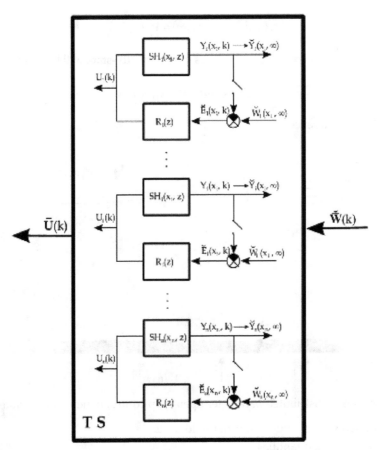

Fig. 9. SISO lumped parameter control loops in the block TS

TS - time part of control synthesis

$\left\{SH_i(x_i,z)\right\}_i$ - time components of HLDS dynamics

$\left\{R_i(z)\right\}_i$ - lumped parameter controllers

$\left\{Y_i\left(x_i,k\right)\right\}_{i,k} / \left\{\breve{W}_i\right\}_i = \left\{\breve{W}_i\left(x_i,\infty\right)\right\}_i$ - controlled/reference quantities

$\left\{\breve{E}_i\left(x_i,k\right)\right\}_{i,k}$ - lumped control errors

$\left\{U_i\left(k\right)\right\}_{i,k}$ - lumped control quantities

$\breve{W}\left(k\right) / \bar{U}\left(k\right)$ - vector of lumped reference/control quantities

When the individual components of the vector $\breve{W} = \left\{\breve{W}_i\left(x_i,\infty\right)\right\}_i$ are input in the particular

control loops: $\left\{SH_i\left(x_i,z\right),R_i\left(z\right)\right\}_i$, the control processes take place. The applied control laws

result in the sequences of control inputs: $\left\{U_i\left(k\right)\right\}_{i,k}$, and respectively the output quantities,

for $k \rightarrow \infty$ converging to reference quantities

$$\left\{Y_i\left(x_i,k\right) \rightarrow Y_i\left(x_i,\infty\right) = \breve{W}_i\left(x_i,\infty\right)\right\}_{i,k} \tag{13}$$

Values of these lumped controlled quantities in new steady-state will be further denoted as

$$\left\{\breve{Y}_i\left(x_i,\infty\right) = \breve{W}_i\left(x_i,\infty\right)\right\}_i \tag{14}$$

see Fig. 9. – 10. for reference.

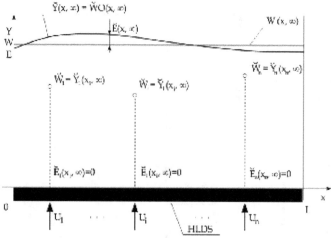

Fig. 10. Quantities of distributed parameter open-loop control in new steady-state

HLDS - controlled system with zero-order hold units

$\left\{U_i\right\}_i$ - lumped control quantities

$\left\{\breve{Y}_i\left(x_i,\infty\right) / \breve{W}_i = \breve{W}_i\left(x_i,\infty\right)\right\}_i$ - controlled/reference quantities in new steady-state

$\left\{\breve{E}_i\left(x_i,\infty\right)\right\}_i$ - lumped control errors

$\breve{Y}\left(x,\infty\right)$ - controlled distributed quantity in new steady-state

$W(x,\infty)$ - distributed reference quantity

$\breve{W}O(x,\infty)$ - unique best approximation of reference quantity

$\breve{E}(x,\infty)$ - distributed control error with minimal norm

Then according to equations (12-14) for the new steady-state it holds: $\left\{\breve{Y}_i(x_i,\infty)\mathcal{H}HR_i(x,\infty)=\breve{W}_i(x_i,\infty)\mathcal{H}HR_i(x,\infty)\right\}_i$, which implies that the overall distributed output quantity at the time $k\to\infty$: $\breve{Y}(x,\infty)$ gives the unique best approximation of the distributed reference variable: $W(x,\infty)$

$$\breve{Y}(x,\infty)=\sum_{i=1}^{n}\breve{Y}_i(x_i,\infty)\mathcal{H}HR_i(x,\infty)=\sum_{i=1}^{n}\breve{W}_i(x_i,\infty)\mathcal{H}HR_i(x,\infty)=\breve{W}O(x,\infty) \tag{15}$$

Therefore the control error has a unique form as well, with minimal quadratic norm

$$\left\|\breve{E}(x,\infty)\right\|=\left\|W(x,\infty)-\breve{Y}(x,\infty)\right\|=\left\|W(x,\infty)-\sum_{i=1}^{n}\breve{Y}_i(x_i,\infty)\mathcal{H}HR_i(x,\infty)\right\|=$$
$$\left\|W(x,\infty)-\sum_{i=1}^{n}\breve{W}_i(x_i,\infty)\mathcal{H}HR_i(x,\infty)\right\|=\left\|W(x,\infty)-\breve{W}O(x,\infty)\right\| \tag{16}$$

Thus the control task, defined at equation (9), is accomplished with $\left\{\breve{E}_i(x_i,\infty)=0\right\}_i$ - see Fig. 10. for reference. As the conclusion of this section we may state that the control synthesis in open-loop control system is realized as:

Time Tasks of Control Synthesis – in lumped parameter control loops

Space Tasks of Control Synthesis – as approximation task.

When mathematical models cannot provide an ideal representation of controlled DPS dynamics and disturbances are present with an overall effect on the output in steady-state expressed by $EY(x,\infty)$ then

$$\left\|W(x,\infty)-\breve{Y}(x,\infty)\pm EY(x,\infty)\right\|\le\left\|W(x,\infty)-\breve{Y}(x,\infty)\right\|+\left\|EY(x,\infty)\right\|=$$
$$\left\|W(x,\infty)-\breve{W}O(x,\infty)\right\|+\left\|EY(x,\infty)\right\|=\left\|\breve{E}(x,\infty)\right\|+\left\|EY(x,\infty)\right\| \tag{17}$$

Finally at the design stage of a control system, for a given desired quality of control δ in space domain, it is necessary to choose appropriate number and layout of actuators for the fulfillment of this requirement

$$\left\|\breve{E}(x,\infty)\right\|+\left\|EY(x,\infty)\right\|\le\delta \tag{18}$$

4.2 Closed-loop control with block RHLDS

Let us consider now a distributed parameter feedback control loop with initial conditions identical as the case above, see Fig. 11. In blocks SS1 a SS2 approximation problems are solved while in block RHLDS reduced distributed output quantities $\left\{YR_i(x,k)\right\}_{i,k}$ are generated. Block TS in Fig. 12., contains the controllers $\{R_i(z)\}_i$ designed as the controllers for SISO lumped parameter control loops $\{SH_i(x_i,z),R_i(z)\}_i$ with respect to the request formulated by equation (12). In the k-th step in block SS2 at approximation of $Y(x,k)$ on the subspace of $\{YR_i(x,k)\}_{i,k}$

$$\min\left\|Y(x,k)-\sum_{i=1}^{n}Y_i(x_i,k)YR_i(x,k)\right\| \tag{19}$$

time components of partial output quantities $\{\breve{Y}_i(x_i,k)\}_i$ are obtained, in block SS1 reference quantities $\{\breve{W}_i(x_i,\infty)\}_i$ are computed. Then on the output of the algebraic block is $\{\breve{E}_i(x_i,k)=\breve{W}_i(x_i,\infty)-\breve{Y}_i(x_i,k)\}_i$. These sequences $\{\breve{E}_i(x_i,k)\}_i$ enter into the TS on $\{R_i(z)\}_i$ and give $\{U_i(k)\}_i$, which enter into HLDS with $\{\breve{Y}_i(x_i,k)\}_i$ on the SS2 output - among $\{U_i(k)\}_i$ and $\{\breve{Y}_i(x_i,k)\}_i$ there are relations $\{SH_i(x_i,z)\}_i$. - This analysis of control synthesis process shows that synthesis in time domain is realized on the level of one parameter control loops $\{SH_i(x_i,z),R_i(z)\}_i$, Fig. 9.

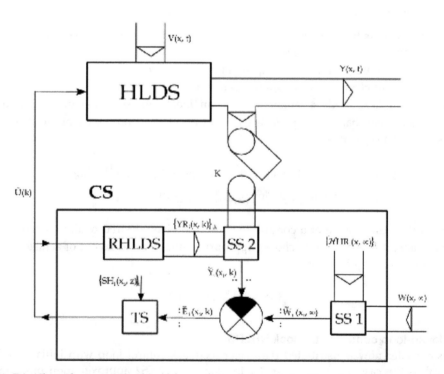

Fig. 11. Distributed parameter closed-loop system of control with reduced space components of output quantity

HLDS - controlled system with zero-order hold units

RHLDS - model for reduced space components: $\{YR_i(x,k)\}_{i,k}$

CS - control synthesis

TS/SS1,SS2 - time/space parts of control synthesis

K - time/space sampling

$Y(x,t)$ - distributed output quantity

$W(x,\infty), V(x,t)$ - reference and disturbance quantities

$\{Y_i(x_i,k)\}_{i,k} = \{\breve{Y}_i(x_i,k)\}_{i,k}$ – time components of output quantity

$\{\breve{W}_i = \breve{W}_i(x_i,\infty)\}_i / \{\breve{E}_i(x_i,k)\}_{i,k}$ - reference quantities/control errors

$\bar{U}(k)$ - vector of lumped control quantities

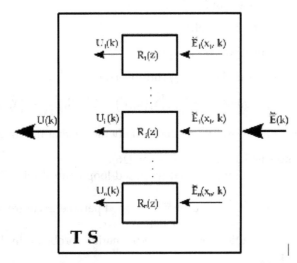

Fig. 12. The block of time synthesis

TS - time part of control synthesis

$\{R_i(z)\}_i$ - lumped parameter controllers

$\breve{\bar{E}}(k) = \{\breve{E}_i(x_i,k)\}_{i,k}$ - vector of lumped control errors

$\bar{U}(k) = \{U_i(k)\}_{i,k}$ - vector of lumped control quantities

For $k \to \infty$ $\{YR_i(x,k) \to \mathcal{H}HR_i(x,\infty)\}_{i,k}$, $\{\breve{Y}_i(x_i,k)\}_{i,k} \to \{\breve{Y}_i(x_i,\infty) = \breve{W}_i(x_i,\infty)\}_i$ along with

$\{\breve{E}_i(x_i,k) \to \breve{E}_i(x_i,\infty) = 0\}_{i,k}$. Thus the control task, defined by equation (9) is accomplished

as given by relation (16). In case of the uncertainty of the control process relations similar to (17-18) are also valid.

Let´s consider now the approximation of $W(x,\infty)$ in the block SS1 in timestep k, on the set of $\{YR_i(x,k)\}_{i,k}$. Then in the control process sequences of quantities $\{\breve{W}_i(x_i,k)\}_{i,k}$ are obtained, as desired quantities of SISO control loops $\{SH_i(x_i,z), R_i(z)\}_i$ which are closed throughout the blocks TS, HLDS and SS2, see Fig. 13.

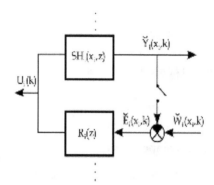

Fig. 13. Lumped parameter SISO control loops – i-th control loop

$SH_i(x_i, z)$ - i-th time component of HLDS dynamics

$R_i(z)$ - i-th lumped parameter controller

$\check{Y}_i(x_i, k) / U_i(k)$ - i-th controlled/control quantity

$\check{W}_i(x_i, k) / \check{E}_i(x_i, k)$ - i-th desired quantity/control error

If for $k \to \infty$ $\left\{ YR_i(x, k) \to \mathcal{HHR}_i(x, \infty) \right\}_{i,k}$, $\left\{ Y_i(x_i, k) = \check{Y}_i(x_i, k) \right\}_{i,k} \to \left\{ \check{Y}_i(x_i, \infty) = \check{W}_i(x_i, \infty) \right\}_i$

along with $\left\{ \check{E}_i(x_i, k) \to \check{E}_i(x_i, \infty) = 0 \right\}_{i,k}$ - this actually means that the control task defined in

equation (9), is accomplished as given by relation (16).

Finally we may state as a summary, that in closed-loop control with RHLDS the control synthesis is realized as:

Time Tasks of Control Synthesis – on the level of lumped parameter control loops

Space Tasks of Control Synthesis – as approximation tasks.

At the same time the solution of the approximation problem in block SS1 on the approximation set $\left\{ YR_i(x, k) \right\}_{i,k}$

$$\min \left\| W(x, \infty) - \sum_{i=1}^{n} W_i(x_i, k) YR_i(x, k) \right\| \tag{20}$$

in timestep k is obtained

$$W(x, \infty) = \sum_{i=1}^{n} \check{W}_i(x_i, k) YR_i(x, k) + \check{E}(x, k) \tag{21}$$

where $\check{E}(x, k)$ is the unique element at the best approximation of $W(x, \infty)$ on the set of approximate functions $\left\{ YR_i(x, k) \right\}_i$. Similary by the solution of approximation problem in the block SS2 - $\left\{ \check{Y}_i(x_i, k) \right\}_i$ in the timestep k distributed output quantity $Y(x, k)$ is given as

$$Y(x, k) = \sum_{i=1}^{n} \check{Y}_i(x_i, k) YR_i(x, k) \tag{22}$$

4.3 Closed-loop control

Let us now consider a distributed parameter feedback loop as featured in Fig. 14. with initial conditions as above, where in timestep k an approximation problem is solved

$$\min \left\| E(x,k) - \sum_{i=1}^{n} E_i(x_i,k) YR_i(x,k) \right\| \tag{23}$$

and as a result in timestep k a vector $\breve{\bar{E}}(k) = \{\breve{E}_i(x_i,k)\}_i$ is obtained. By relations (20-22) the further equations are valid

$$
\begin{aligned}
&\min \left\| E(x,k) - \sum_{i=1}^{n} E_i(x_i,k) YR_i(x,k) \right\| = \\
&= \min \left\| W(x,\infty) - Y(x,k) - \sum_{i=1}^{n} E_i(x_i,k) YR_i(x,k) \right\| = \\
&= \min \left\| W(x,\infty) - \sum_{i=1}^{n} \left[\breve{Y}_i(x_i,k) + E_i(x_i,k) \right] YR_i(x,k) \right\| = \\
&= \left\| W(x,\infty) - \sum_{i=1}^{n} \left[\breve{Y}_i(x_i,k) + \breve{E}_i(x_i,k) \right] YR_i(x,k) \right\|
\end{aligned}
\tag{24}
$$

The problem solution $W(x,\infty) = \sum_{i=1}^{n} \left[\breve{Y}_i(x_i,k) + \breve{E}_i(x_i,k) \right] YR_i(x,k) + \breve{E}(x,k)$ is obtained by approximation.

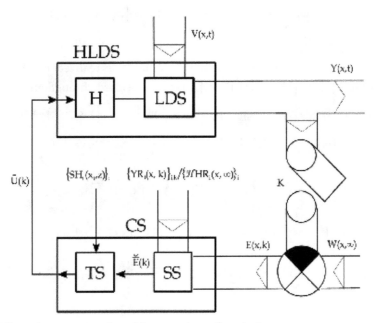

Fig. 14. Distributed parameter closed-loop system of control

HLDS - LDS with zero-order hold units

CS – control synthesis
TS /SS – time/space control synthesis
K – time/space sampling
$Y(x,t)/W(x,\infty)$ – distributed controlled/reference quantity
$\{SH_i(x_i,z)\}_i$ - transfer functions - dynamic characteristics of HLDS in time domain
$\{YR_i(x,k)\}_i / \{\mathcal{H}HR_i(x,\infty)\}_i$ - reduced characteristics in space domain
$E(x,k)$ – distributed control error
$V(x,t)$ – distributed disturbance quantity
$\breve{\bar{E}}(k)$ – vector of control errors
$\bar{U}(k)$ – vector of control quantities
Comparison of relation (21) and   result of the approximation problem (24)   gives

$$\sum_{i=1}^{n} \breve{W}_i(x_i,k)YR_i(x,k) + \breve{E}(x,k) = \sum_{i=1}^{n}\left[\breve{Y}_i(x_i,k) + \breve{E}_i(x_i,k)\right]YR_i(x,k) + \breve{E}(x,k) \qquad \text{and} \qquad \text{then}$$

$$\sum_{i=1}^{n} \breve{W}_i(x_i,k)YR_i(x,k) = \sum_{i=1}^{n}\left[\breve{Y}_i(x_i,k) + \breve{E}_i(x_i,k)\right]YR_i(x,k), \qquad\qquad \text{finally}$$

$\left\{\breve{W}_i(x_i,k) = \breve{Y}_i(x_i,k) + \breve{E}_i(x_i,k)\right\}_{i,k}$ is obtained. Now in vector form this means

$$\breve{\bar{W}}(k) = \breve{\bar{Y}}(k) + \breve{\bar{E}}(k) \Rightarrow \breve{\bar{E}}(k) = \breve{\bar{W}}(k) - \breve{\bar{Y}}(k) = \left\{\breve{E}_i(x_i,k) = \breve{W}_i(x_i,k) - \breve{Y}_i(x_i,k)\right\}_{i,k} \qquad (25)$$

Then sequences $\left\{\breve{E}_i(x_i,k)\right\}_{i,k}$ enter into the block TS on $\{R_i(z)\}_i$ and give $\{U_i(k)\}_i$, among $\{U_i(k)\}_i$ and $\left\{\breve{Y}_i(x_i,k)\right\}_i$ there are relations $\{SH_i(x_i,z)\}_i$. Finally this analysis of control synthesis process shows that synthesis in time domain is realized on the level of one parameter control loops $\{SH_i(x_i,z),R_i(z)\}_i$, Fig. 13. - closed throughout the structure of distributed parameter control loop, Fig. 14. If for $k \to \infty$ $\{YR_i(x,k) \to \mathcal{H}HR_i(x,\infty)\}_i$ and $\left\{\breve{Y}_i(x_i,k)\right\}_{i,k} \to \left\{\breve{Y}_i(x_i,\infty) = \breve{W}_i(x_i,\infty)\right\}_i$ along with $\left\{\breve{E}_i(x_i,k) \to \breve{E}_i(x_i,\infty) = 0\right\}_{i,k}$ then

$$\min\left\|E(x,\infty) - \sum_{i=1}^{n}E_i(x_i,\infty)\mathcal{H}HR_i(x,\infty)\right\| = \left\|W(x,\infty) - \breve{Y}(x,\infty) - \sum_{i=1}^{n}\breve{E}_i(x_i,\infty)\mathcal{H}HR_i(x,\infty)\right\| =$$

$$= \left\|\sum_{i=1}^{n}\breve{W}_i(x_i,\infty)\mathcal{H}HR_i(x,\infty) + \breve{E}(x,\infty) - \sum_{i=1}^{n}\breve{Y}_i(x_i,\infty)\mathcal{H}HR_i(x,\infty) - \sum_{i=1}^{n}\breve{E}_i(x_i,\infty)\mathcal{H}HR_i(x,\infty)\right\| = (26)$$

$$= \left\|\sum_{i=1}^{n}\left[\breve{W}_i(x_i,\infty) - \breve{Y}_i(x_i,\infty)\right]\mathcal{H}HR_i(x,\infty) + \breve{E}(x,\infty) - \sum_{i=1}^{n}\breve{E}_i(x_i,\infty)\mathcal{H}HR_i(x,\infty)\right\| = \left\|\breve{E}(x,\infty)\right\|$$

is valid, so the above given control task (9) is accomplished - whereas in the steady-state $\left\{\breve{W}_i(x_i,\infty) - \breve{Y}_i(x_i,\infty) = \breve{E}_i(x_i,\infty) = 0\right\}_i$. By concluding the above presented discussion, the control synthesis in closed-loop control is realized as:

Time Tasks of Control Synthesis – on the level of lumped parameter control loops
Space Tasks of Control Synthesis – as approximation tasks.

When mathematical models cannot provide an ideal description of controlled DPS dynamics and disturbances are present with an overall effect on the output in steady-state, expressed by $EY(x,\infty)$ then the realtions similar to (17-18) are also valid here.

In practice mostly only reduced distributed parameter transient responses in steady-state $\{\mathcal{H}HR_i(x,\infty)\}_i$ are considered for the solution of the approximation tasks in the block SS of

the scheme in Fig. 14. along with robustification of controllers $\{R_i(z)\}_i$.

For simplicity problems of DPS control have been formulated here for the distributed desired quantity $W(x,\infty)$. In case of $W(x,k)$ is assumed, the control synthesis is realized similarly:

- *In Space Domain* - as problem of approximation in particular sampling intervals
- *In Time Domain* - as control synthesis in lumped parameter control loops, closed throughout structures of the distributed parameter control loop.

The solution of the presented problems of control synthesis an assumption is used, that in the framework of the chosen control systems the prescribed control quality can be reached both in the in space and time domain. However in the design of actual control systems for the given distributed parameter systems, usually the

• optimization of the number and layout of actuators
• optimization of dynamical characteristics of lumped/distributed parameter actuators
• optimization of dynamical characteristics of lumped parameter control loops

is required and necessary.

5. Distributed Parameter Systems Blockset for MATLAB & Simulink

As a software support for DPS modelling, control and design of problems in MATLAB & Simulink the programming environment **Distributed Parameter Systems Blockset for MATLAB & Simulink (DPS Blockset)** - a Third-Party Product of The MathWorks www.mathworks.com/products/connections/ – Fig. 15., has been developed within the program CONNECTIONS of The MathWorks Corporation by the Institute of Automation, Measurement and Applied Informatics of Mechanical Engineering Faculty, Slovak University of Technology in Bratislava (IAMAI-MEF-STU) (Hulkó et al., 2003-2010). Fig. 16. shows The library of **DPS Blockset**. The **HLDS** and **RHLDS** blocks model controlled DPS dynamics described by numerical structures as LDS with zero-order hold units - H. **DPS Control Synthesis** provides feedback to distributed parameter controlled systems in control loops with blocks for discrete-time **PID, Algebraic, State-Space and Robust Synthesis**. The block **DPS Input** generates distributed quantities, which can be used as distributed reference quantities or distributed disturbances, etc. **DPS Display** presents distributed quantities with many options including export to AVI files. The block **DPS Space Synthesis** performs space synthesis as an approximation problem.

As a demonstration, some results of the discrete-time PID control of complex-shape metal body heating by the **DPS Blockset** are shown in Fig. 17.-19., where the heating process was modelled by finite element method in the COMSOL Multiphysics virtual software environment - www.comsol.com.

The block **Tutorial** presents methodological framework for formulation and solution of control problems for DPS. The block **Show** contains motivation examples such as: *Control of temperature field of 3D metal body* (the controlled system was modelled in the virtual software environment COMSOL Multiphysics); *Control of 3D beam of „smart" structure* (the controlled system was modelled in the virtual software environment ANSYS); *Adaptive control of glass furnace* (the controlled system was modelled by Partial Differential Equations Toolbox of the MATLAB), and *Groundwater remediation control* (the controlled system was modelled in the virtual software environment MODFLOW). The block **Demos** contains examples oriented at the methodology of modelling and control synthesis. The **DPS Wizard** gives an automatized guide for arrangement and setting distributed parameter control loops in step-by-step operation.

Fig. 15. Distributed Parameter Systems Blockset on the web portal of The MathWorks Corporation

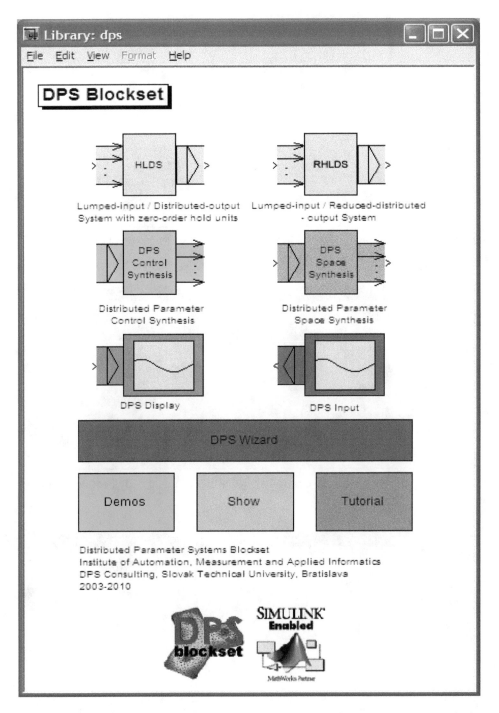

Fig. 16. The library of Distributed Parameter Systems Blockset for MATLAB & Simulink –
Third-Party Product of The MathWorks

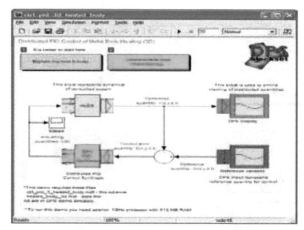

Fig. 17. Distributed parameter control loop for discrete-time PID control of heating of metal body in DPS Blockset environment

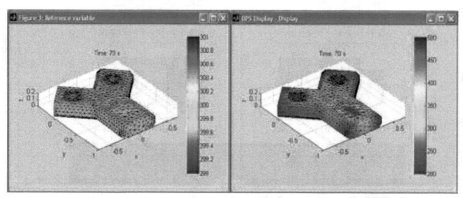

Fig. 18. Distributed reference and controlled quantities of metal body heating over the numerical net

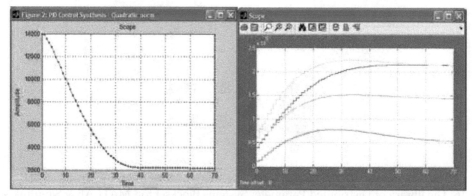

Fig. 19. Quadratic norm of distributed control error and discrete lumped actuating quantities at discrete-time PID control of heating of metal body in DPS Blockset environment

6. Interactive control via the Internet

For the interactive formulation and solution of DPS demonstration control problems via the Internet, an **Interactive Control** service has been started on the web portal **Distributed Parameter Systems Control** - www.dpscontrol.sk of the IAMAI-MEF-STU (Hulkó, 2003-2010) – see Fig. 20. for a screenshot of the site. In the framework of the problem formulation, first the computational geometry and mesh are chosen in the complex 3D shape definition domain, then DT distributed transient responses are computed in virtual software environments for numerical dynamical analysis of machines and processes. Finally, the distributed reference quantity is specified in points of the computational mesh - Fig. 18. Representing the solution to those interested animated results of actuating quantities, quadratic norm of control error, distributed reference and controlled quantity are sent in the form of **DPS Blockset** outputs – see Fig. 17-19. for illustration.

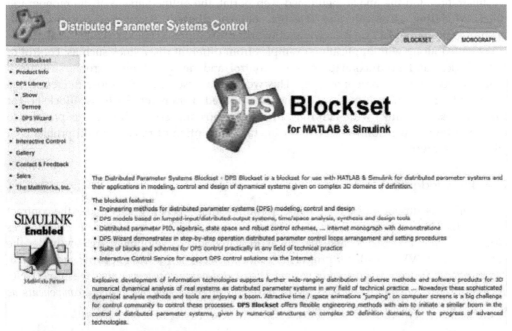

Fig. 20. Web portal Distributed Parameter Systems Control with monograph Modeling, Control and Design of Distributed Parameter Systems with Demonstrations in MATLAB and service Interactive Control

7. Conclusion

The aim of this chapter is to present a philosophy of the engineering approach for the control of DPS – given by numerical structures, which opens a wide space for novel applications of the toolboxes and blocksets of the MATLAB & Simulink software environment. This approach is based on the general decomposition into time and space components of controlled DPS dynamics represented by numerically computed distributed parameter transient and impulse characteristics, given on complex shape definition domains in 3D. Starting out from this dynamics decomposition a methodical framework is presented for the analogous decomposition of control synthesis into the space and time subtasks. In space domain approximation problems are solved, while in the time domain control synthesis is realized by lumped parameter SISO control loops (Hulkó et al., 1981-2010). Based on these decomposition a software product named **Distributed Parameter Systems Blockset for MATLAB & Simulink** - a Third-Party software product of The MathWorks - www.mathworks.com/products/connections/ has been developed within the program CONNECTIONS of The MathWorks Corporation, (Hulkó et al., 2003-2010), where time domain toolboxes and blocksets of software environment MATLAB & Simulink as Control Systems Toolbox, Simulink Control Design, System Identification Toolbox,... are made use of. In the space domain approximation problems are solved as optimization problems by means of the Optimization Toolbox.

For the further support of research in this area a web portal named **Distributed Parameter Systems Control** - www.dpscontrol.sk was realized (Hulkó et al., 2003-2010), see Fig. 20. for an illustration. On the above mentioned web portal, the online version of the monograph titled **Modeling, Control and Design of Distributed Parameter Systems with Demonstrations in MATLAB** - www.mathworks.com/support/books/ (Hulkó et al., 1998), is presented along with application examples from different disciplines such as: control of technological and production processes, control and design of mechatronic structures, groundwater remediation control, etc. This web portal also offers for those interested the download of the demo version of the **Distributed Parameter Systems Blockset for MATLAB & Simulink** with **Tutorial** , **Show** , **Demos** and **DPS Wizard**. This portal also offers the **Interactive Control** service for interactive solution of model control problems of DPS via the Internet.

8. Acknowledgment

This work was supported by the Slovak Scientific Grant Agency VEGA under the contract No. 1/0138/11 for project *„Control of dynamical systems given by numerical structures as distributed parameter systems"* and the Slovak Research and Development Agency under the contract No. APVV-0160-07 for project *„Advanced Methods for Modeling, Control and Design of Mechatronical Systems as Lumped-input and Distributed-output Systems"* also the project No. APVV-0131-10 *„High-tech solutions for technological processes and mechatronic components as controlled distributed parameter systems"*.

9. References

Butkovskij, A. G. (1965). *Optimal control of distributed parameter systems.* Nauka, Moscow
        (in Russian)
Hulkó, G. et al. (1981). On Adaptive Control of Distributed Parameter Systems, *Proceedings of 8-th World Congress of IFAC*, Kyoto, 1981
Hulkó, G. et al. (1987). Control of Distributed Parameter Systems by means of Multi-Input and Multi-Distributed-Output Systems, *Proceedings of 10-th World Congress of IFAC*, Munich, 1987
Hulkó, G. (1989). Identification of Lumped Input and Distributed Output Systems, *Proceedings of 5-th IFAC / IMACS / IFIP Symposium on Control of Distributed Parameter Systems*, Perpignan, 1989

Hulkó, G. et al. (1990). Computer Aided Design of Distributed Parameter Systems of Control, *Proceedings of 11-th World Congress of IFAC*, Tallin, 1990

Hulkó G. (1991). Lumped Input and Distributed Ouptut Systems at the Control of Distributed Parameter Systems. *Problems of Control and Information Theory*, Vol. 20, No. 2, pp. 113-128, Pergamon Press, Oxford

Hulkó, G. et al. (1998). *Modeling, Control and Design of Distributed Parameter Systems with Demonstrations in MATLAB*, Publishing House STU, ISBN 80-227-1083-0, Bratislava

Hulkó, G. et al. (2005). Web-based control design environment for distributed parameter systems control education, *Proceedings of 16-th World Congress of IFAC*, Prague, 2005

Hulkó, G. et al. (2003-2010). *Distributed Parameter Systems Control*. Web portal, Available from: www.dpscontrol.sk

Hulkó, G. et al. (2003-2010). *Distributed Parameter Systems Blockset for MATLAB & Simulink*, www.mathworks.com/products/connections/ - Third-Party Product of The MathWorks, Bratislava-Natick, Available from: www.dpscontrol.sk

Hulkó, G. et al. (2009). Engineering Methods and Software Support for Modeling and Design of Discrete-time Control of Distributed Parameter Systems, *Mini-tutorial, Proceedings of European Control Conference 2009*, Budapest, 2009

Hulkó, G. et al. (2009). Engineering Methods and Software Support for Modelling and Design of Discrete-time Control of Distributed Parameter Systems. *European Journal of Control*, Vol. 15, No. Iss. 3-4, *Fundamental Issues in Control*, (May-August 2009), pp. 407-417, ISSN 0947-3580

Hulkó, G. et. al (2010). Control of Technological Processes Modelled in COMSOL Multiphysics as Distributed Parameter Systems, *Proceedings of Asian COMSOL Conference*, Bangalore, 2010

Lasiecka, I., Triggiani, R. (2000). *Control Theory for Partial Differential Equations* (Encyclopedia of Mathematics and Its Applications 74), Cambridge U. Press, Cambridge, UK

Lions, J. L. (1971). *Optimal control of systems governed by partial differential equations*, Springer-Verlag, Berlin - Heidelberg - New York

Shadrin, A. (2005). *Approximation theory*. DAMTP University of Cambridge, Cambridge UK, Available from: www.damtp.cam.ac.uk

Wang, P. K. C. (1964). *Control of distributed parameter systems* (Advances in Control Systems: Theory and Applications, 1.), Academic Press, New York

3

Linearization of Permanent Magnet Synchronous Motor Using MATLAB and Simulink

A. K. Parvathy and R. Devanathan
Hindustan Institute of Technology and Science, Chennai
India

1. Introduction

Permanent magnet machines, particularly at low power range, are widely used in the industry because of their high efficiency. They have gained popularity in variable frequency drive applications. The merits of the machine are elimination of field copper loss, higher power density, lower rotor inertia and a robust construction of the rotor (Bose 2002).

In order to find effective ways of designing a controller for PM synchronous motor (PMSM), the dynamic model of the machine is normally used. The dynamic model of PM motor can be derived from the voltage equations referred to direct (d) and quadrature (q) axes (Bose 2002).The model derived essentially has quadratic nonlinearity. Linear control techniques generally fail to produce the desired performance. Feedback linearization is a technique that has been used to control nonlinear systems effectively.

By applying exact linearization technique (Cardoso & Schnitman 2011) it is possible to linearize a system and apply linear control methods. But this requires that certain system distributions have involutive property. An approximate feedback linearization technique was formulated by Krener (Krener 1984) based on Taylor series expansion of distributions for non-involutive systems.

Chiasson and Bodson (Chiasson & Bodson 1998) have designed a controller for electric motors using differential geometric method of nonlinear control based on exact feedback linearization. But from a practical point of view, this technique suffers from singularity issues. If the system goes into a state, during the course of the system operation, where the singularity condition is satisfied, then the designed controller will fail.

Starting with the quadratic model of PMSM, we apply quadratic linearization technique based on coordinate and state feedback. The linearization technique used is the control input analog of Poincare's work (Arnold 1983) as proposed by Kang and Krener (Kang & Krener 1992) and further developed by Devanathan (Devanathan 2001,2004) .The quadratic linearization technique proposed is on the lines of approximate linearization of Krener (Krener 1984) and does not introduce any singularities in the system compared to the exact linearization methods reported in (Chiasson & Bodson 1998).

MATLAB simulation is used to verify the effectiveness of the linearization technique proposed. In this chapter, MATLAB/SIMULINK modeling is used to verify the effectiveness of the quadratic linearization technique proposed. In particular, the application of MATLAB and SIMULINK as tools for simulating the following is described:

i.   Dynamic model of a sinusoidal Permanent Magnet Synchronous machine(PMSM)
ii.   Application of nonlinear coordinate and state feedback transformations to linearize the PMSM model and
iii. Tuning the transformations against a linear system model put in Brunovsky form employing error back propogation.

As these applications were somewhat sophisticated , customization and improvisation of the MATLAB/SIMULINK tools were essential. These applications are described in detail in this chapter.

In section 2, linearization of dynamic model of PMSM is discussed and simulation results using MATLAB are given. In section 3, linearization of PMSM machine model is given. Construction of PMSM model using SIMULINK and verification of linearization of PMSM SIMULINK model is given in Section 4. Tuning of the linearizing transformations to account for unmodelled dynamics is discussed in Section 5. In section 6, the chapter is concluded.

2. Linearization of dynamic model of PMSM

2.1 Dynamic model of PMSM

The dynamic model of a sinusoidal PM machine, considering the flux-linkage λ_f to be constant and ignoring the core-loss, can be written as (Bose 2002).

$$\dot{i}_d = \frac{v_d}{L_d} - \frac{R}{L_d}i_d + \frac{L_q}{L_d}p\omega_r i_q \tag{1}$$

$$\dot{i}_q = \frac{v_q}{L_q} - \frac{R}{L_q}i_q - \frac{L_d}{L_q}p\omega_r i_d - \frac{\lambda_f p\omega_r}{L_q} \tag{2}$$

$$\dot{\omega}_r = \frac{1.5p}{J}[\lambda_f i_q + (L_d - L_q)i_d i_q] \tag{3}$$

where all quantities in the rotor reference frame are referred to the stator.

L_q, L_d are q and d axis inductances respectively; R is the resistance of the stator windings; i_q, i_d are q and d axis currents respectively; v_q, v_d are q and d axis voltages respectively; ω_r is the angular velocity of the rotor; λ_f is the amplitude of the flux induced by the permanent magnets of the rotor in the stator phases and p is the number of pole pairs.

Using linear coordinate and state feedback transformations (Kuo 2001) the dynamic model can be written with the linear part put in Brunovsky form (Parvathy et. al. 2005,2006) as

$$\dot{x} = Ax + Bu + f^{(2)}(x) \tag{4}$$

where u $=[\ u_1 \quad u_2\ ]^T = [\ v_q \quad v_d\ ]^T ; x = [i_q \quad i_d \quad \omega_e]^T$

$$A = \begin{bmatrix} 0 & 1 & 0 \\ 0 & 0 & 1 \\ 0 & 0 & 0 \end{bmatrix} \quad B = \begin{bmatrix} 0 & 0 \\ 0 & 1 \\ 1 & 0 \end{bmatrix}; \ f^{(2)}(x) = \begin{bmatrix} b_1 x_2 x_3 \\ b_2 x_1 x_3 \\ b_3 x_1 x_2 \end{bmatrix}$$

where b_1, b_2, b_3 are constants derived from the motor parameters.

2.2 Linearization of the dynamic model

Coordinate and state feedback transformations in quadratic form (Kang & Krener 1992)

$$y = x + \phi^{(2)}(x) \tag{5}$$

$$u = (I_2 + \beta^{(1)}(x))v + \alpha^{(2)}(x) \tag{6}$$

Where

$$y = \begin{bmatrix} y_1 & y_2 & y_3 \end{bmatrix}^T ; v = \begin{bmatrix} v_1 & v_2 \end{bmatrix}^T$$

$\phi^{(2)}(x); \alpha^{(2)}(x)$ and $\beta^{(1)}(x)$ are derived by solving the Generalized Homological Equations (Kang & Krener 1992) .Applying the transformations (5) and (6), (4) is reduced to

$$\dot{y} = Ay + Bv + O^{(3)}(y,v)$$

where $O^{(3)}(y,v)$ represent third and higher order nonlinearities .

2.3 Verification of linearization using MATLAB function

The problem now is to apply MATLAB to verify the theoretical result on quadratic linearization of the dynamic model (4). Expanding (4), we can see that the expression of the derivative of each state variable has the other two state variables in it. This becomes difficult to solve using manual methods of differential equation solution. The tool selected for solving the dynamic equations is the MATLAB function called ODE45.

ODE45 is a MATLAB function that solves initial value problems for ordinary differential equations (ODEs). It uses the iterative Runge Kutta method of solving equations. Hence, this function does not return the solution as an expression, but the values of the solution function at discrete instants of time. ODE45 is based on an explicit Runge-Kutta formula, the Dormand-Prince pair. It is a *one-step* ode45 –in the sense that, in computing $y(t_n)$, it needs only the solution at the immediately preceding time point $y(t_{n-1})$ .In general, ode45 is the best function to apply as a "first try" for most problems.

[t,Y] = ode45(odefun,tspan,y0) with tspan = [t0 tf] integrates the system of differential equations from time t0 to tf with initial conditions y0. Function f = odefun(t,y) is defined, where t corresponds to the column vector of time points and y is the solution array. Each row in y corresponds to the solution at a time returned in the corresponding row of t. To obtain solutions at the specific times t0, t1,...,tf (all increasing or all decreasing), we use tspan = [t0,t1,...,tf].

[t, Y] = ode45(odefun,tspan,y0,options) solves as above with default integration parameters replaced by property values specified in 'options'. Commonly used properties include a scalar relative error tolerance RelTol (1e-3 by default) and a vector of absolute error tolerances AbsTol (all components are 1e-6 by default).

The PLOT function is used to create a computer-graphic plot of any two quantities or a group of quantities with respect to time.

A simulation of (4) was carried out for different values of inputs u_1 and u_2 in the open loop before applying the linearizing transformations where b_1 = -0.0165*(10^-3) ; b_2 -186.96; b_3 =5754386. Fig 1 shows a plot of x_3 (angular velocity) versus time for pulse inputs

$u_1 = 0.1\{u(t)-u(t-1)\}$ and $u_2=0.1\{u(t)-u(t-1)\}$ where $u(t)$ is a unit step function. The system is oscillatory as seen from Fig 1. Fig. 2 shows the result of simulation of equation (4) after the linearizing transformations (5) and (6) are applied to the system. It shows a plot of y_1, y_2, y_3 against time for pulse inputs $v_1=0.2\{u(t)-u(t-1)\}$, $v_2=0.2\{u(t)-u(t-1)\}$. From Fig 2 it is seen that the system shows a stable response for a pulse input in v_1 and v_2 even under open loop conditions.

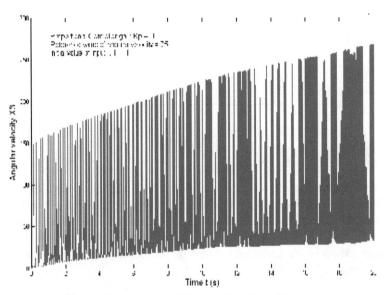

Fig. 1. Time response of angular velocity with $v_1=0.1$ and $v_2=0.1$

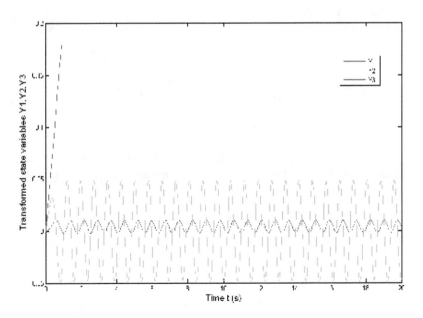

Fig. 2. Time response of Y_1 , Y_2 , Y_3 with $v_1=0.2$ and $v_2=0.2$

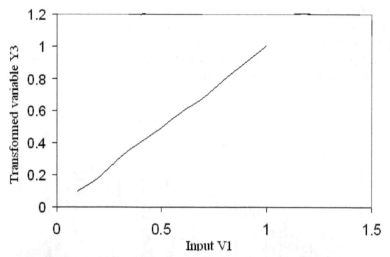

Fig. 3. Variation of transformed variable Y₃ with input v₁ (keeping v₂=0.1)

Fig. 3 shows the steady state gain of y_3 with respect to v_1 while v_2 is maintained constant. It is observed that the plot between y_3 and v_1 is almost linear, thus verifying that the linearizing system is almost linear. A similar test before linearization revealed that the steady gain of x_3 with respect to input u_1 varied over a large range thus revealing the nonlinearity.
Thus the linearization of the dynamic model of PMSM was verified through the use of MATLAB function ODE45 .

3. Linearization of PMSM machine model

3.1 PMSM model in normal form
The 4 – dimensional PM machine model can be derived as below (Bose 2002).

$$\dot{x} = Ax + Bu + f^{(2)}(x)$$
$$x = \begin{bmatrix} x_1 & x_2 & x_3 & x_4 \end{bmatrix}^T = \begin{bmatrix} \theta & \omega_e, & i_q, & i_d \end{bmatrix}^T ; \tag{7}$$
$$u = \begin{bmatrix} u_1 & u_2 \end{bmatrix}^T = \begin{bmatrix} v_q & v_d \end{bmatrix}^T$$

where v_q, v_d, i_q, i_d represent the quadrature and direct axis voltages and currents respectively and θ, ω_e represent rotor position and angular velocity respectively.

$$A = \begin{bmatrix} 0 & 1 & 0 & 0 \\ 0 & 0 & \dfrac{1.5p\lambda}{J} & 0 \\ 0 & \dfrac{-\lambda p}{L_q} & \dfrac{-R}{L_q} & 0 \\ 0 & 0 & 0 & \dfrac{-R}{L_d} \end{bmatrix} ; B = \begin{bmatrix} 0 & 0 \\ 0 & 0 \\ \dfrac{1}{L_q} & 0 \\ 0 & \dfrac{1}{L_d} \end{bmatrix} ; f^{(2)}(x) = \begin{bmatrix} 0 \\ \dfrac{1.5p(L_d - L_q)i_d i_q}{J} \\ \dfrac{-L_d p\omega_e i_d}{L_q} \\ \dfrac{L_q p\omega_e i_q}{L_d} \end{bmatrix}$$

λ is the flux induced by the permanent magnet of the rotor in the stator phases. L_d, L_q are the direct and quadrature inductances respectively. R is the stator resistance. p is the number of pole pairs and J is the system moment of inertia.

The model (7) can be reduced to normal form for two inputs (Brunovsky 1970), in a standard way using the following transformations (Kuo 2001),

$$x = \begin{bmatrix} a_1c_1 & 0 & 0 & 0 \\ 0 & a_1c_1 & 0 & 0 \\ 0 & 0 & c_1 & 0 \\ 0 & 0 & 0 & a_4 \end{bmatrix} y$$

$$u = \begin{bmatrix} 1 & 0 \\ 0 & a_4/c_2 \end{bmatrix} u' + \begin{bmatrix} 0 & -a_1a_2 & -a_3 & 0 \\ 0 & 0 & 0 & -a_4^2/c_2 \end{bmatrix} y$$

(8)

where

$$a_1 = \frac{1.5p\lambda}{J}; \; a_2 = \frac{-\lambda p}{L_q}, a_3 = \frac{-R}{L_q}, \; a_4 = \frac{-R}{L_d}; \; c_1 = \frac{1}{L_q}; c_2 = \frac{1}{L_d}$$

The Brunovsky form for two inputs is given below (where x, u, A and B are retained for simplicity of notation).

$$\dot{x} = Ax + Bu + f^{(2)}(x)$$

(9)

Where

$$A = \begin{bmatrix} 0 & 1 & 0 & 0 \\ 0 & 0 & 1 & 0 \\ 0 & 0 & 0 & 0 \\ 0 & 0 & 0 & 0 \end{bmatrix}; B = \begin{bmatrix} 0 & 0 \\ 0 & 0 \\ 1 & 0 \\ 0 & 1 \end{bmatrix}; f^{(2)}(x) = \begin{bmatrix} 0 \\ k_1 x_3 x_4 \\ k_2 x_2 x_4 \\ k_3 x_2 x_3 \end{bmatrix}$$

$$k_1 = \frac{1.5p(L_d - L_q)a_4}{Ja_1}, k_2 = \frac{-L_d pa_1 a_4}{L_q}, k_3 = \frac{L_q C_1^2 a_1}{L_d a_4}$$

3.2 Linearization of PMSM normal form model

Given the 4 dimensional model of a PM synchronous motor (IPM model) of the form (9), the system can be linearized using the following transformations

$$y = x + \phi^{(2)}(x)$$

(10)

$$u = (I_2 + \beta^{(1)}(x))v + \alpha^{(2)}(x)$$

(11)

where $u = [u_1 \quad u_2]^T; v = [v_1 \quad v_2]^T$

$$\phi^{(2)}(x) = \begin{bmatrix} 0 \\ 0 \\ k_1 x_3 x_4 \\ 0 \end{bmatrix}, \alpha^{(2)}(x) = \begin{bmatrix} -k_2 x_2 x_4 \\ -k_3 x_2 x_3 \end{bmatrix}, \beta^{(1)}(x) = -\begin{bmatrix} k_1 x_4 & k_1 x_3 \\ 0 & 0 \end{bmatrix}$$

where I_2 is the identity matrix of order 2. The system then reduces to

$$\dot{y} = Ay + Bv + O^{(3)}(y,v)$$

where $O^{(3)}(y,v)$ represents third and higher order nonlinearities .

4. Linearization of PMSM model using SIMULINK

4.1 Construction of PMSM model
Customization of PMSM machine model using MATLAB/SIMULINK tools had to be carried out as the standard library available contained only special cases. The PM motor drive simulation was built in several steps through the construction of q-axis circuit, d-a xis circuit, torque block and speed block.

4.1.1 q-axis circuit
By using the following system equation the q-axis circuit is constructed.

$$v_q = R_q i_q + \omega_r (L_d i_d + \lambda_f) + \rho L_q i_q$$

q-axis circuit in the SIMULINK is shown in Fig.4 .

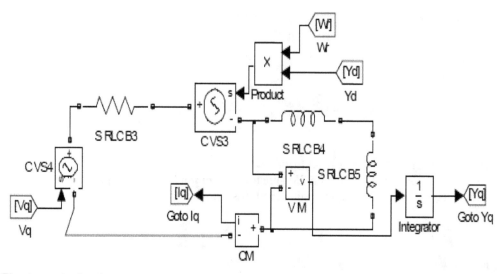

Fig. 4. q-axis circuit

4.1.2 d-axis circuit

By using the following system equation the d-axis circuit is constructed.

$$v_d = R_d i_d + \omega_r (L_q i_q) + \rho(\lambda_f + L_d i_d)$$

Simulation of the d-axis circuit is shown in Fig. 5.

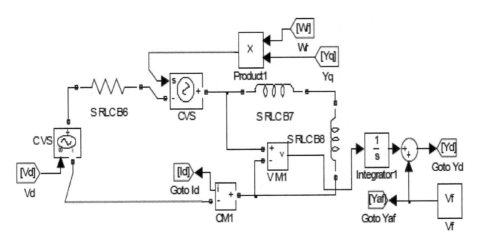

Fig. 5. d-axis circuit

4.1.3 Torque block

By using the following torque equation the torque block is constructed.

$$T_e = (3/2)(p/4)(\lambda_d i_q - \lambda_q i_d)$$

The simulation of torque T_e circuit is shown in the Fig 6.

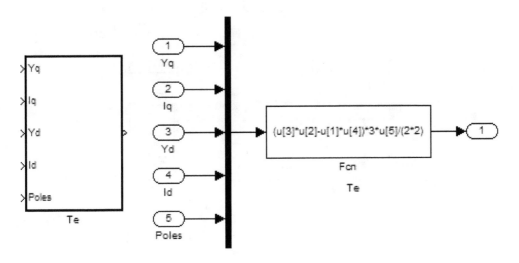

Fig. 6. T_e Block in Simulink

4.1.4 Speed block

By using the following equations the speed block is constructed.

$$\omega_m = \int (T_e - T_l - B\omega_m)/J dt$$

$$\omega_m = \omega_r (2/p)$$

The simulation of ω_m circuit is shown in the Fig.7.

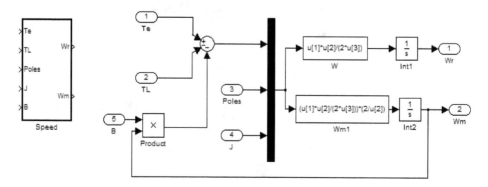

Fig. 7. Speed Block in Simulink

4.1.5 Designed PMSM with V$_q$ and V$_d$ as inputs

The PMSM is constructed by using q-axis, d-axis, torque and speed blocks (figures 4,5, 6, and 7) and is shown in figure 8.

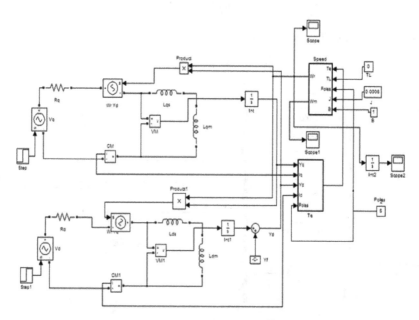

Fig. 8. Permanent Magnet Synchronous Motor with V_q and V_d as inputs

4.2 Linearization of PMSM SIMULINK model

The PMSM model is first converted to controller normal form of the linear part (9) using transformations as given in (8). Then linearization of the PMSM is carried out using quadratic coordinate and state feedback as given in (10) and (11). Fig.9 gives the PMSM model including linearization blocks. N_1 and N_2 include the linear transformations (8) while L_1 and L_2, denote the nonlinear transformations (11) and (10) respectively. N_1, N_2, L_1 and L_2 are implemented using function blocks.

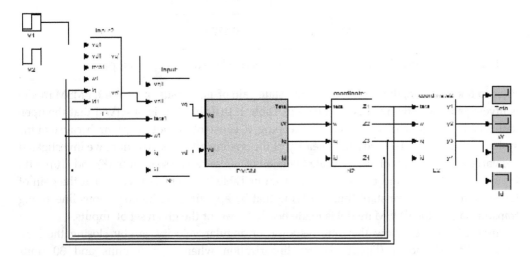

Fig. 9. Linearization of PMSM model

4.3 Verification of linearization of PMSM-simulation results

For the Interior PMSM, parameters are taken as follows:

Stator resistance R = 2.875Ω; q- axis Inductance L_q = 9mH; d-axis Inductance L_d = 7mH; Flux induced in magnets λ_f = 0.175 wb; Moment of Inertia J = 0.0008 kg.m^2 Friction factor B = 1 N.m.s; No. of pole pairs p = 4

v_q	ω_e	K=dω_e/dv_q
5	2.55	-
10	4.7709	0.44418
15	6.4706	0.33994
20	7.6082	0.22752
25	8.2567	0.1297
30	8.5326	0.05518

Table 1. Steady state gain of ω_e versus v_q for the system in open loop (Prior to linearization)

v_1	y_2 *10^(-6)	K=d y_2 /d v_1 *10^(-6)
5	2.176	-
10	4.366	0.438
15	6.585	0.4438
20	8.845	0.452
25	11.162	0.4634
30	13.55	0.4776

Table 2. Steady state gain of y_2 vs v_1 for the linearized system in open loop

Prior to linearization, the open loop steady state gain of ω_e versus v_q of the PMSM model is investigated and the results are given in Table 1. In Table 1, it is observed that the open loop steady state gain of ω_e versus v_q (keeping v_d constant) is not constant because of the system nonlinearity. To verify the linearity of the system after linearization, we investigated the variation of its gain of y_2 (a scaled version of ω_e as can be seen from (8) and (10)) with input v_1 (see Fig. 9) and the results are given in Table 2. The table reveals that the gain of the system is nearly constant thus verifying that by applying the homogeneous linearizing transformation, the PMSM model is made nearly linear for the given set of inputs.

Figures 10 and 11 show the time response of angular velocity ω_e by closing the loop around PMSM model (Fig. 8) before linearization when v_q = 5 units and 30 units respectively. It is observed that the dynamic response for v_q = 5 is more oscillatory compared to the case of v_q = 30 with a fixed controller of proportional gain = 50 and integral constant =2. This is to be expected since the loop gain is higher in the former case with a higher static gain in the plant or motor as can be seen from Table I.

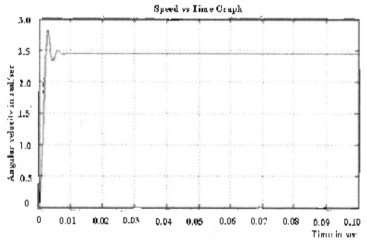

Fig. 10. Time response of angular velocity in closed loop when v_q = 5; k_p = 50; k_i = 2 (before linearization)

Figures 12 and 13 show the time response of y_2 of the transformed PMSM system (Fig. 9) in closed loop when $v_1 = 5$ units and 30 units respectively . It is observed that a uniform output response is obtained in the closed loop after linearization when the reference is varied. Since the static gain in Table II is nearly uniform, the loop gain is also nearly constant for the extreme points in the operating range, thus resulting in the uniform dynamic responses in Fig. 12 and 13.

Simulation results show that the nearly constant gain of the linearized model, results in a uniform response on a range of set point and load inputs with a fixed controller. This is in contrast to the case before linearization under the corresponding conditions.

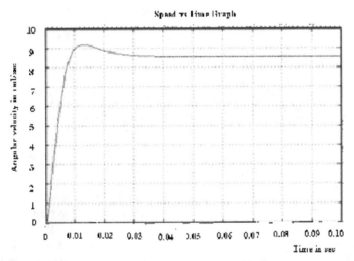

Fig. 11. Time response of angular velocity in closed loop when $v_q = 30$; $k_p = 50$; $k_i = 2$ (before linearization)

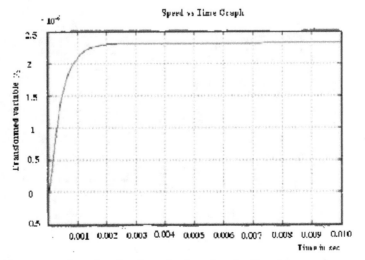

Fig. 12. Time Response of y_2 for the linearized system in closed loop when $v_1 = 5$; $k_p = 50$; $k_i = 2$

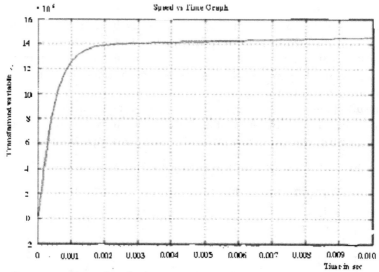

Fig. 13. Time Response of y_2 for the linearized system in closed loop when $v_1 = 30$; $k_p = 50; k_i = 2$

5. Tuning linearizing transformations

Unmodelled dynamics coupled with the third and higher order nonlinearities introduced due to quadratic linearization, are best accounted for by tuning the transformations (Levin & Narendra 1993).

To further improve the linearity of the system taking into account unmodelled dynamics and higher order nonlinearities, tuning of the transformation parameters against an actual PM machine is done on the lines similar to those proposed by Narendra (Levin & Narendra 1993). Fig 14 shows the block diagram for tuning.

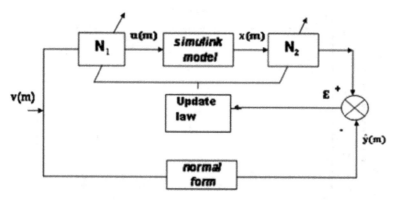

Fig. 14. Block diagram for tuning of transformation

Referring to Fig. 14, error (E) can be calculated as

$$E = (\varepsilon^T \varepsilon) = \left[(y - \hat{y})^T (y - \hat{y}) \right]^{1/2} \qquad (12)$$

where $\varepsilon = \begin{pmatrix} \varepsilon_1 & \varepsilon_2 & \varepsilon_3 & \varepsilon_4 \end{pmatrix}^T$, $\varepsilon_i = y_i - \hat{y}_i ; i = 1,2,3,4$

The error can be written as

$$E = \left(\varepsilon_1^2 + \varepsilon_2^2 + \varepsilon_3^2 + \varepsilon_4^2 \right)^{1/2}$$

Since $\phi^{(2)}(x)$ and $\beta^{(1)}(x)$ are both functions of k_1, we shall redefine

$$\phi^{(2)}(x) = \begin{bmatrix} 0 \\ 0 \\ k_1 x_3 x_4 \\ 0 \end{bmatrix}$$

and

$$\beta^{(1)}(x) = - \begin{bmatrix} k_1' x_4 & k_1' x_3 \\ 0 & 0 \end{bmatrix}$$

so that $\phi^{(2)}(x)$ and $\beta^{(1)}(x)$ can be independently tuned by tuning k_1 and k_1' respectively and $\alpha^{(2)}(x)$ is not varied.

5.1 Updation of N_2 transformation coefficients

Tuning of N_2 transformation implies the tuning of $\phi^{(2)}(x)$. As $\phi^{(2)}(x)$ is a function of only $k_1 x_3 x_4$, the coefficient k_1 has to be updated based on the error between the outputs of quadratic linearized system and normal form. The updation law is derived as follows.

$$\Delta k_1 = \frac{\partial E}{\partial k_1} = \frac{\partial E}{\partial y} \frac{\partial y}{\partial k_1}$$

From (12), it is seen that

$$\frac{\partial E}{\partial y_i} = \frac{\varepsilon_i}{E} ; i = 1,2,3,4$$

Hence

$$\frac{\partial E}{\partial y} = \begin{bmatrix} \dfrac{\varepsilon_1}{E} & \dfrac{\varepsilon_2}{E} & \dfrac{\varepsilon_3}{E} & \dfrac{\varepsilon_4}{E} \end{bmatrix}.$$

$$\therefore \Delta k_1 = \begin{bmatrix} \dfrac{\varepsilon_1}{E} & \dfrac{\varepsilon_2}{E} & \dfrac{\varepsilon_3}{E} & \dfrac{\varepsilon_4}{E} \end{bmatrix} \begin{bmatrix} 0 \\ 0 \\ x_3 x_4 \\ 0 \end{bmatrix} = \frac{\varepsilon_3}{E} x_3 x_4 \qquad (13)$$

Updation of $\phi^{(2)}(x)$ is done by using the formula:

$$k_1(m) = k_1(m-1) - \rho\Delta k_1(m); 0 < \rho < 1 \tag{14}$$

where m corresponds to the updating step and ρ correspond to the accelerating factor.

5.2 Updation of N_1 transformation coefficients

Tuning of N_1, transformation is achieved by tuning of $\beta^{(1)}(x)$. As $\beta^{(1)}(x)$ is a function of $k_1'x_3$ and $k_1'x_4$, the coefficient k_1' has to be updated based on the error between the outputs of quadratic linearized system and normal form . The updation law is derived as follows.

$$\Delta k_1' = \frac{\partial E}{\partial k_1'} = \frac{\partial E}{\partial y}\frac{\partial y}{\partial x}\frac{\partial x}{\partial u_1}\frac{\partial u_1}{\partial k_1'}$$

where ; $\dfrac{\partial E}{\partial y} = \begin{bmatrix} \dfrac{\varepsilon_1}{E} & \dfrac{\varepsilon_2}{E} & \dfrac{\varepsilon_3}{E} & \dfrac{\varepsilon_4}{E} \end{bmatrix}$

$$\frac{\partial y}{\partial x} = \begin{bmatrix} 1 & 0 & 0 & 0 \\ 0 & 1 & 0 & 0 \\ 0 & 0 & (1+k_1'x_4) & k_1'x_3 \\ 0 & 0 & 0 & 1 \end{bmatrix}; \frac{\partial x}{\partial u_1} = \begin{bmatrix} 0 \\ -\dfrac{1}{k_2x_4} \\ 0 \\ -\dfrac{1}{k_2x_2} \end{bmatrix}$$

Assuming that the steady state of the Simulink model is reached within the tuning period.

$$\frac{\partial u_1}{\partial k_1'} = -v_1x_4 - v_2x_3$$

Hence

$$\Delta k_1' = \frac{(v_1x_4 + v_2x_3)}{E}\left(\frac{\varepsilon_2}{k_2x_4} + \frac{\varepsilon_3k_1'x_3 + \varepsilon_4}{k_2x_2}\right) \tag{15}$$

Tuning of the quadratic linearizing transformations is done by updating the transformation coefficients of vector polynomials $\phi^{(2)}(x)$ and $\beta^{(1)}(x)$.

5.3 Construction of controller tuning blocks

Updation of $\phi^{(2)}(x)$ is done by using (14) and Δk_1 can be obtained from (13).

The tuning is done using Memory blocks and they are constructed in simulink as shown. Fig. 15 shows the construction of Del k_1 block. Similar construction can be done for Del k_1' where $\Delta k_1'$ can be obtained from (14).

Simulation of updation of $\phi^{(2)}(x)$ and $\beta^{(1)}(x)$ can be done using the simulation diagrams Fig. 15, 16 and 17.

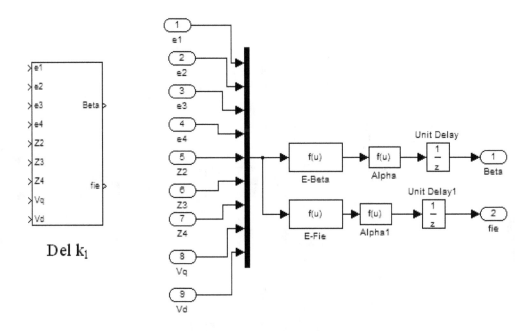

Fig. 15. Calculation of del K_1 for the updation of $\phi^{(2)}(x)$

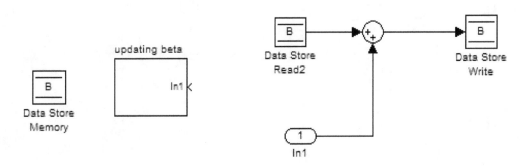

Fig. 16. Updation rule using Memory Read and memory Write Blocks

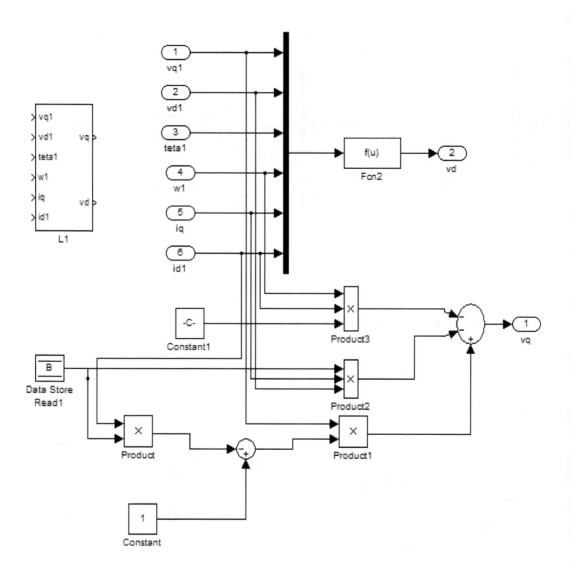

Fig. 17. Updation of input transformation

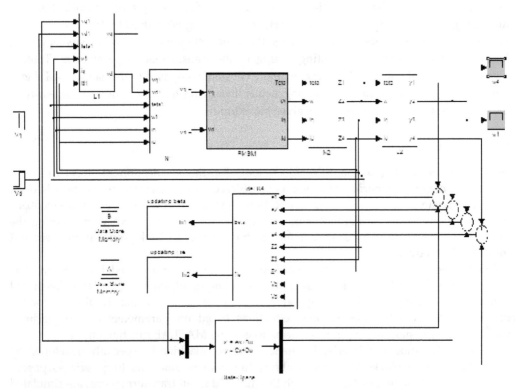

Fig. 18. Simulation diagram including controller tuning

v_1	$y_2 *10^{-6}$	$K=d\,y_2\,/d\,v_1$ $*10^{-6}$
106	7.142	--
108	7.38	0.048
110	7.334	0.048
112	7.43	0.048
114	7.525	0.0475
116	7.62	0.0475
118	7.714	0.047

Table 3. Steady state gain of y_2 versus v_1 for the linearized system in open loop after tuning

Complete simulation diagram including conversion to Brunovsky form, linearization and tuning is given in Fig.18. It is seen that the error after tuning is reduced to 0.01.

In Table 3, improvements are obtained for the steady state gain of y_2 versus v_1 for the linearized system after incorporating tuning of the transformation parameters. Table 3 reveals that the gain of the system is even more constant compared to that shown in Table 2, thus verifying that by tuning the homogeneous linearizing transformation, the linearity of the system has been improved for the given set of inputs.

6. Conclusion

Application of MATLAB and SIMULINK tools for the verification of linearization of permanent magnet synchronous motor is considered in this chapter. Simulation is done using the dynamic model of PMSM, application of nonlinear coordinate and state feedback transformations to the SIMULINK model which is customized to PMSM and tuning the transformations against a linear system model employing error back propogation to account for unmodelled dynamics.

Initially, linearization of PMSM is verified using the dynamic model of PMSM. The dynamic model of a PM synchronous motor involving quadratic nonlinearity is linearized and simulated using MATLAB. Steps are given to perform dynamic simulation for the nonlinear system using the dynamic equations based on parameters of the machine, together with the state and input transformations using MATLAB function ODE45.

The SIMULINK model of Interior Permanent Magnet machine is specially developed by integrating various blocks as standard library functions do not cater to generic purposes . The PMSM machine model, together with the state and input transformations, are simulated using SIMULINK. The simulation results show that the linearizing transformations effectively linearize the system thus supporting the theory.

To account for the unmodelled dynamics and third and higher order nonlinearities, tuning of the transformation parameters is done by comparing the output of the linearized system with a normal form output. Tuning the transformation functions $\phi^{(2)}(x)$ and $\beta^{(1)}(x)$ is shown further to improve the linearity of the resulting system.

More details of the simulation results are given in (Parvathy et. al. 2011).

7. Acknowledgement

The authors acknowledge the financial support rendered by the management of Hindustan Institute of Technology and Science, Chennai in publishing this chapter.

8. References

Bimal .K. Bose(2002) *Modern Power Electronics and Ac Drives* , Pearson Education.

Gildeberto S. Cardoso & Leizer Schnitman (2011) Analysis of exact linearization and approximate feedback linearization techniques.
          www.hindawi.com/journals/mpe/aip/205939.pdf

Krener A.J. (1984) Approximate linearization by state feedback and coordinate change. *Systems and control Letters.* Vol. 5,pp. 181-185.

John Chiasson& Marc Bodson (1998) .Differential – Geometric methods for control of lectric motors. *International Journal of Robust and Nonlinear Control*. No.8, pp.923-954.

Arnold V.I.(1983) *Geometric methods in the theory of ordinary differential equations*. Springer-Verlag, NewYork, pp 177-188.

Kang .W & Krener.A.J. (1992) . Extended quadratic controller normal form and dynamic state feedback linearisation of nonlinear systems. *SIAM Journal of Control and optimization* . No.30,pp.1319-1337.

Devanathan R. (2001) . Linearisation Condition through State Feedback. *IEEE transactions on Automatic Control* .ol 46,no.8,pp.1257-1260

Devanathan R. (2004). Necessary and sufficient conditions for quadratic linearisation of a linearly controllable system. *INT.J. Control*, Vol.77, No.7,pp. 613-621

Benjamin.C.Kuo(2001) . *Automatic Control Systems*. Prentice-Hall India.

A.K.Parvathy, Aruna Rajan, R.Devanathan (2005) . Complete Quadratic Linearization of PM Synchronous Motor Model. , *proceedings of NEPC conference*. IIT Karagpur, pp 49-52.

A.K.Parvathy, R.Devanathan (2006) Linearisation of Permanent Magnet Synchronous Motor Model.*proc.of IEEE ICIT 2006*,pp. 483-486.

Pavol Brunovsky (1970) *A classification of linear controllable systems*. Kybernetika,Vol. 6. No.3, pp. 173-188

Asriel U Levin & Kumpati S Narendra (1993). Control of nonlinear dynamical systems using neural networks: controllability and stabilization. *IEEE Transactions on neural networks*.Vol-4, No.2,1993, pp.192-206.

A.K.Parvathy, V.Kamaraj & R.Devanathan (2011) A generalized quadratic linearization technique for PMSM. accepted for publication in , *European Journal of Scientific Research*.

4

Thermal Behavior of IGBT Module for EV (Electric Vehicle)

Mohamed Amine Fakhfakh, Moez Ayadi,
Ibrahim Ben Salah and Rafik Neji
University of Sfax/Sfax
Tunisia

1. Introduction

EVs are divided into three categories: the pure EV, the hybrid EV, and the fuel cell EV. Although these three types of electric vehicle have different system configuration, one (or more) motor drive system is always needed to convert electrical power into mechanical ones. Among the drive systems used for EV, induction motor system and permanent magnet motor systems are mostly used for their high power density, high efficiency.

The motor drive system for electric vehicle (EV) is composed of a battery, three phase inverter, a permanent magnet motor, and a sensor system. The inverter is a key unit important among these electrical components which converts the direct current of the battery into the alternating current to rotate the motor. Therefore, for predicting the dynamic power loss and junction temperature, the electro-thermal coupling simulation techniques to estimate the power loss and to calculate the junction temperature become important.

This paper describes a compact thermal model suitable for the electro-thermal coupling simulation of EV inverter module for two current control methods. We can predict the dynamic temperature rise of Si devices by simulating the inverter operation in accordance with the real EV running.

2. Dynamic model of the EV

As shown in Figure 1 and table 1, there are six forces acting on the electric vehicle: the rolling resistance force, the aerodynamic force, the aerodynamic lift force, the gravity force, the normal force, and the motor force.

2.1 Rolling resistance force

Rolling resistance is due the tires deforming when contacting the surface of a road and varies depending on the surface being driven on. It can be model using the following equation:

$$F_1 = f\,M_v\,g \tag{1}$$

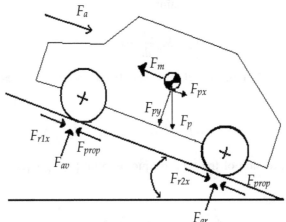

Fig. 1. Diagram of forces applied to the EV

F_{r1x}	Rolling resistance force
F_{r2x}	Rolling resistance force
F_{av}	Normal force
F_{ar}	Normal force
F_a	Aerodynamic force
F_{prop}	Thrust force
F_p	Gravity force
F_m	Motor force
θ	Slope angle with the horizontal

Table 1. Applied forces to EV

2.2 Aerodynamic force

Aerodynamic drag is caused by the momentum loss of air particles as they flow over the hood of the vehicle. The aerodynamic drag of a vehicle can be modeled using the following equation:

$$F_2 = \frac{1}{2}\rho S_f C_x V^2 \tag{2}$$

2.3 Gravity force

The gravity force can be calculated as follows:

$$F_3 = M_v g \sin\theta \tag{3}$$

2.4 Motor force

Using Newton's Second Law, we can deduce the motor force; it can be obtained by the following equation:

$$M_v \frac{dV}{dt} = \sum \vec{F}_{ext} = \vec{F}_m + \vec{F}_p + \vec{F}_a + \vec{F}_r \tag{4}$$

By projection on the (O, x) axis, we obtain:

$$F_m = M_v \frac{dV}{dt} + F_a + F_p + F_r \tag{5}$$

The power that the EV must develop at stabilized speed is expressed by the following equation:

$$P_{vehicle} = F_m V = (F_r + F_a + F_p + M_v \frac{dV}{dt})V \tag{6}$$

We deduce the expression of the total torque by multiplying equation (5) with the wheel radius R:

$$C_{vehicle} = C_r + C_a + C_p + M_v R \frac{dV}{dt} \tag{7}$$

Neglecting the mechanical losses in the gearbox, the t electromagnetic torque C_{em} developed by the motor is obtained by dividing the wheels torque $C_{vehicle}$ by the ratio reduction r_d.

$$C_{em} = \frac{1}{r_d}\left(C_r + C_a + C_p + M_v R \frac{dV}{dt} \right) \tag{8}$$

Figure 2 presents the dynamic model of the EV load, implemented under Matlab/simulink.

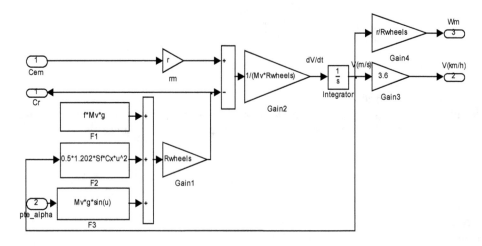

Fig. 2. SIMULINK dynamic model of electric vehicle

3. Electric motor control

Control of permanent magnet synchronous motor is performed using field oriented control. The stator windings of the motor are fed by an inverter that generates a variable frequency variable voltage. The frequency and phase of the output wave are controlled using a position sensor as shown in figure 3.

In our studie, we have used two types of current control, Hysteresis and PWM.

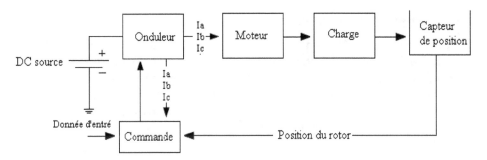

Fig. 3. Drive system schematic

3.1 PWM current controller
PWM current controllers are widely used. The switching frequency is usually kept constant. They are based in the principle of comparing a triangular carrier wave of desire switching frequency and is compared with error of the controlled signal [Bose, 1996].

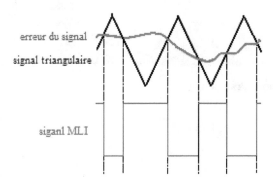

Fig. 4. PWM current controller

3.2 Hysteresis current controller
Hysteresis current controller can also be implemented to control the inverter currents. The controller will generate the reference currents with the inverter within a range which is fixed by the width of the band gap [Bose, 1996; Pillay et al., 1989].

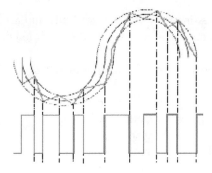

Fig. 5. Hysteresis current controller

4. Thermal model of IGBT module

The studied module is the Semikron module SKM 75GB 123D (75A/1200V) which contains two IGBTs and with two antiparallel diodes. The structure of the module contains primarily eight layers of different materials, each one of it is characterized by its thickness Li, its thermal conductivity Ki, density ρi and its heat capacity Cpi. Table 2 show the materials properties of the various layers of module as shown in figure 6. These values are given by the manufacturer and/or of the literatures [Dorkel et al., 1996; Uta et al., 2000; Thoams et al., 2000].

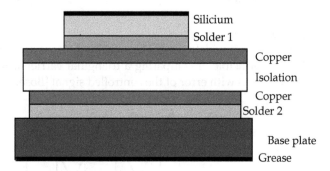

Fig. 6. Example of the module structure

Material	L (mm)	K (W/mK)	ρCp (J/Kcm³)
Silicium	0.4	140	1.7
Solder 1	0.053	35	1.3
Copper	0.35	360	3.5
Isolation	0.636	100	2.3
Copper	0.35	360	3.5
Solder 2	0.103	35	1.3
Base plate	3	280	3.6
Grease	0.1	1	2.1

Table 2. Thermal parameters of a power module

In the power module, the heating flow diffuses vertically and also laterally from the heating source. So, a thermal interaction happens inside the module between the adjacent devices when they operate together.

This thermal interaction depends from [Kojima et al., 2006; Ayadi et al., 2010; Fakhfakh et al., 2010]:
- The dissipated power value of the various components.
- The disposition of the chip components.
- The boundary condition at the heat spreader.

Figure 7 shows the thermal influence between the different components of the module. We notice that each component has a thermal interaction with the others and we supposed that each module have zero interaction with other modules.

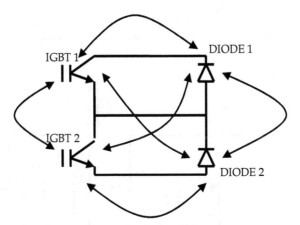

Fig. 7. Different thermal influences between the module components

Literature proposes some thermal circuit networks for electrothermal simulation for the semiconductor device. For example the finite difference method (FDM) and the finite element method (FEM). In our study we have used the FEM technique to model our inverter module. Figure 8 shows the thermal circuit example obtained by the FEM of IGBT1 without thermal interaction.

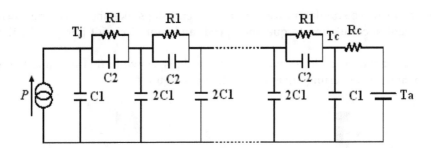

Fig. 8. Thermal circuit obtained by the FEM

Where:
- P is the input power dissipation device.
- Tj is the junction temperature.
- R1 is the thermal resistance.
- Rc is the convection resistance.
- C1 and C2 are thermal capacitance.
- Ta is the ambient temperature.

In order to introduce the thermal interaction between the different components of the module, we inserted three other current sources P1, P2 and P3. These sources are deduced from the structure of IGBT module [Drofenik et al., 2005; Hamada et al., 2006; Usui et al., 2006].

The source P1 is the power loss of DIODE1; it is introduced at the interface between the silicon and the copper materials because the IGBT1 and the DIODE1 ships are bounded on the same copper area. The source P2 and P3 are power loss of IGBT2 and DIODE2, they are introduced between solder 2 and base plate because all module components have the base plate as a common material. So the thermal circuit network of IGBT1 becomes as the figure 9.

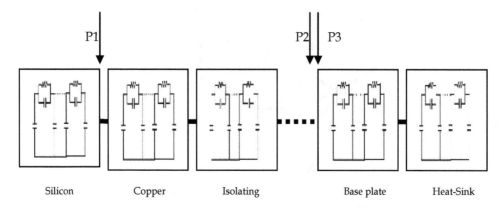

Fig. 9. Thermal model of IGBT module

5. Simulation and results

The PM motor drive simulation was built in several steps like abc phase transformation to dqo variables, calculation torque and speed, and control circuit [Ong, 1998; Roisse et al., 1998].

Parks transformation used for converting Iabc to Idq is shown in figure 10 and the reverse transformation for converting Idq to Iabc is shown in figure 11.

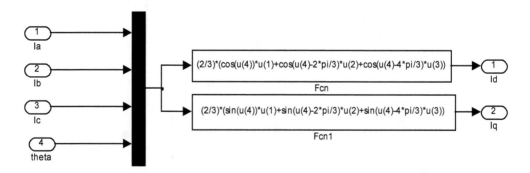

Fig. 10. Iabc to Idq bloc

The inverter is implemented in Simulink as shown in figure 12. The inverter consists of the "universal bridge" with the parameters of the IGBT module studied. All the voltages and the currents in the motor and the inverter can be deducted. The following figure shows the model of the inverter used.

For proper control of the inverter using the reference currents, current controllers are implemented generate the gate pulses for the IGBT's. Current controllers used are shown in figure 13 and 14.

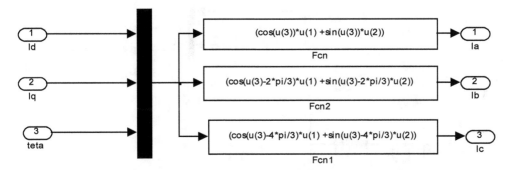

Fig. 11. Idq to Iabc bloc

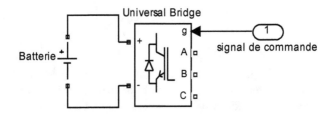

Fig. 12. Inverter model

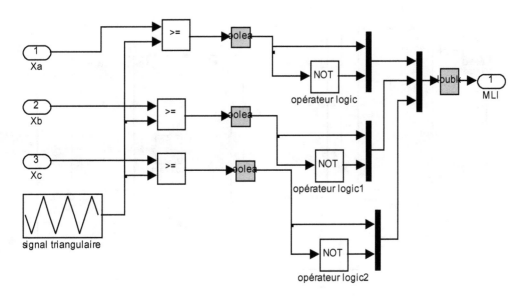

Fig. 13. PWM current controller

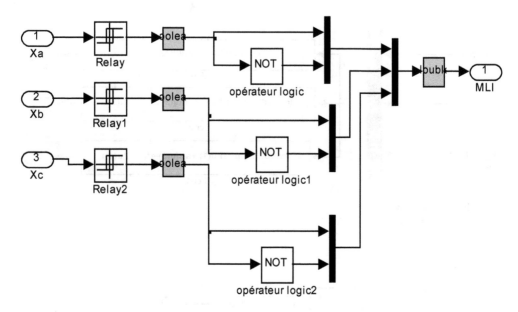

Fig. 14. Hysteresis controller

The complete system used for simulation and implemented in MATLAB / Simulink, is shown in Figure 15. This system was tested with two current controls, hysteresis and PWM control. The motor used is an axial flux Permanent Magnet Synchronous Motor (PMSM). For the simulation, we controlled the speed of EV at 30km / h.

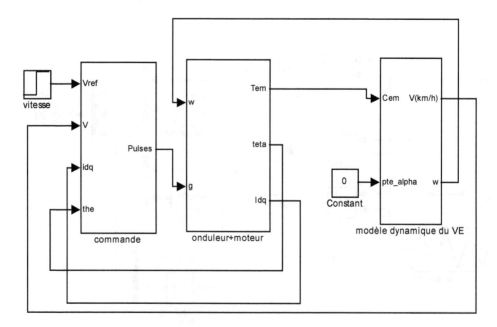

Fig. 15. PMSM in a traction chain

Figure 16 shows the EV speed regulated at 30km / h for the two types of control. We note that with the hysteresis control, we reach faster the steady state.

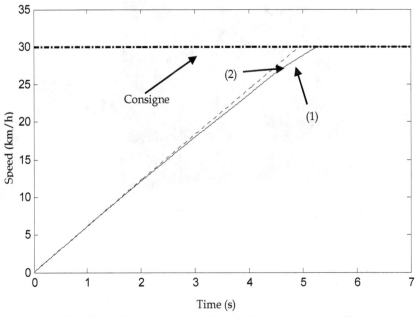

Fig. 16. EV speed; (1): with PWM controller; (2): with hysteresis controller

The stator phase currents corresponding to this regulation are represented by figure 17 and 18 Figure 19 and 20 show the IGBT1 and DIODE1 power losses for hysteresis and PWM current control respectively.

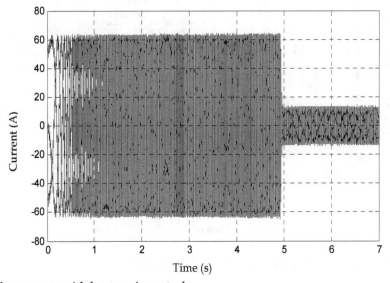

Fig. 17. Iabc currents with hysteresis control

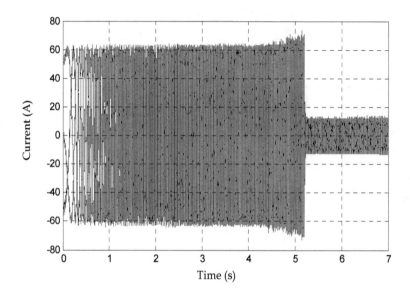

Fig. 18. Iabc currents with PWM control

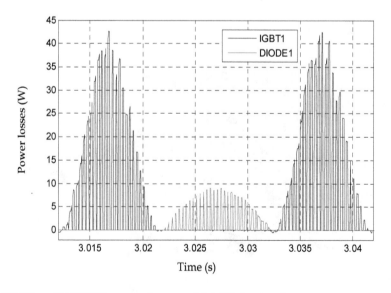

Fig. 19. IGBT1 and DIODE1 power losses with PWM control

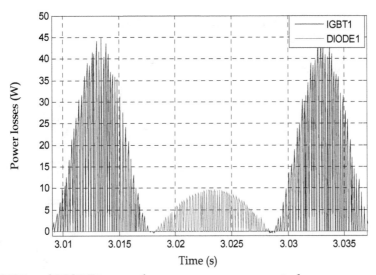

Fig. 20. IGBT1 and DIODE1 power losses with hysteresis control

Figure 21 and 22 show the IGBT1 and DIODE1 junction temperature obtained by the two types of current control. It is very clear that the junction temperature of IGBT1 and DIODE1 is higher for the hysteresis control; this is due by the increase of power dissipation of the module components this type of control.

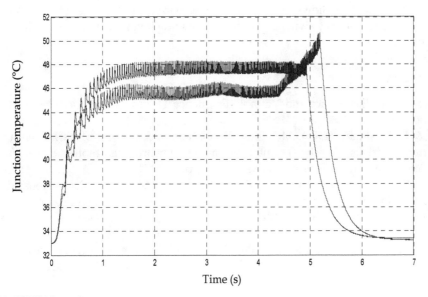

Fig. 21. IGBT1 junction temperature

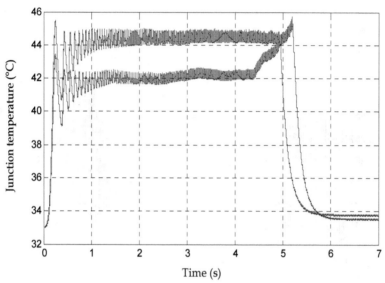

Fig. 22. DIODE1 junction temperature

6. Conclusion

A detailed dynamic model for EV was studied using two current control systems. MATLAB / Simulink were chosen from several simulation tools because of its flexibility in working with analog and digital devices, it is able to represent real-time results with the simulation time reduced. A comparative study was carried out in terms of switching frequency for power dissipated by the components of the inverter and junction temperature. The hysteresis current control has a variable switching frequency that depends on the hysteresis band, this type of control allows for fast simulations with a shorter time. The PWM current control has a fixed frequency switching and allows having junction temperatures lower than the hysteresis control.

7. References

B. K. Bose, Power Electronics and Variable Frequency Drives. (1996). 1 ed: Wiley, John & Sons

P. Pillay & R. Krishnan. (1989). Modeling, simulation, and analysis of permanent-magnet motor drives. I. The permanent-magnet synchronous motor drive. *Industry Applications, IEEE Transactions on*, vol. 25, pp. 265-273

Jean-Marie Dorkel, Patrick Tounsi, & Philippe Leturcq. (1996). Three-Dimensional thermal Modeling Based on the Two-Port Network Theory for Hybrid or Monolithic Integrated Power Circuits. IEEE Transaction on Electronics Devices, vol. 19, NO. 4, pp. 501-507

Uta Hecht & Uwe Scheuermann. (2000). Static and Transient Thermal Resistance of Advanced Power Modules. *Semikron Elektronik GmbH*, Sigmundstr. 200, 90431 Nürnberg (Germany).

Thomas Stockmeier. (2000). Power semiconductor packaging-a problem or a resource? From the state of the art to future trends. *Semikron Elektronik GmbH*, Sigmundstr. 2000, 90431 Nürnberg (Germany)

M. Ayadi, M.A. Fakhfakh, M. Ghariani, & R. Neji. (2010). Electrothermal modeling of hybrid power modules. *Emerald, Microelectronics International (MI)*, volume 27, issue 3, 2010, pp. 170-177

M.A. Fakhfakh, M. Ayadi, and R. Neji. Thermal behavior of a three phase inverter for EV (Electric Vehicle). *in Proc of 15th IEEE Mediterranean Electromechanical Conference (MELECON'10)*, Valletta, Malta, April 25-28, 2010, C4P-E24-3465, pp.1494-1498.

Kojima, et al. Novel Electro-thermal Coupling Simulation Technique for Dynamic Analysis of HV (Hybrid Vehicle) Inverter. *Proceedings of PESCO6*, pp. 2048-2052, 2006.

Hamada. Novel Electro-Thermal Coupling Simulation Technique for Dynamic Analysis of HV (Hybrid Vehicle) Inverter," Proc. of 7thIEEE Power Electronics Specialists Conference (PESC 2006), pp.2048-2052, 2006

U. Drofenik & J. Kolar. (2005). A Thermal Model of a Forced-Cooled Heat Sink for Transient Temperature Calculations Employing a Circuit Simulator. *Proceedings of IPECNiigata* 2005, pp. 1169-1177, 2005

M. Usui, M. Ishiko, "Simple Approach of Heat Dissipation Design for Inverter Module," Proc. of International Power Electronics Conference (IPEC 2005), pp. 1598-1603, 2005.

C. M. Ong. (1998). Dynamic simulation of electric machinery using MATLAB/Simulink.

H. Roisse, M. Hecquet, P. Brochet. (1998). Simulation of synchronous machines using a electric-magnetic coupled network model. IEEE Trans. on Magneticss, vol.34, pp.3656-3659, 1998.

5

Design and Simulation of Legged Walking Robots in MATLAB® Environment

Conghui Liang, Marco Ceccarelli and Giuseppe Carbone
LARM: Laboratory of Robotics and Mechatronics, University of Cassino
Italy

1. Introduction

It is well known that legged locomotion is more efficient, speedy, and versatile than the one by track and wheeled vehicles when it operates in a rough terrain or in unconstructed environment. The potential advantages of legged locomotion can be indicated such as better mobility, obstacles overcoming ability, active suspension, energy efficiency, and achievable speed (Song & Waldron, 1989). Legged walking robots have found wide application areas such as in military tasks, inspection of nuclear power plants, surveillance, planetary explorations, and in forestry and agricultural tasks (Carbone & Ceccarelli, 2005; González et al., 2006; Kajita & Espiau, 2008).

In the past decades, an extensive research has been focused on legged walking robots. A lot of prototypes such as biped robots, quadrupeds, hexapods, and multi-legged walking robots have been built in academic laboratories and companies (Kajita & Espiau, 2008). Significant examples can be indicated as ASIMO (Sakagami et al., 2002), Bigdog (Raibert, 2008), Rhex (Buehler, 2002), and ATHLETE (Wilcox et al., 2007). However, it is still far away to anticipate that legged walking robots can work in a complex environment and accomplish different tasks successfully. Mechanical design, dynamical walking control, walking pattern generation, and motion planning are still challenge problems for developing a reliable legged walking robot, which can operate in different terrains and environments with speedy, efficient, and versatility features.

Mechanism design, analysis, and optimization, as well as kinematic and dynamic simulation of legged walking robots are important issues for building an efficient, robust, and reliable legged walking robot. In particular, leg mechanism is a crucial part of a legged walking robot. A leg mechanism will not only determine the DOF (degree of freedom) of a robot, but also actuation system efficiency and its control strategy. Additionally, it is well understood that a torso plays an important role during animal and human movements. Thus, the aforementioned two aspects must be taken into account at the same time for developing legged walking robots.

Computer aided design and simulation can be considered useful for developing legged walking robots. Several commercial simulation software packages are available for performing modeling, kinematic, and dynamic simulation of legged walking robots. In particular, Matlab® is a widely used software package. It integrates computation, visualization, and programming in an easy-to-use environment where problems and solutions are expressed in familiar mathematical notation. By using a flexible programming environment, embedded functions, and several useful simulink® toolboxes, it is relative easy and fast to perform kinematic and dynamic analysis of a robotic mechanical system (Matlab manual, 2007). Additionally, motion control and task planning algorithms can be tested for a proposed mechanism design before implementing them on a prototype.

In this chapter, the applications of Matlab® tool for design and simulation of legged walking robots are illustrated through three cases, namely a single DOF biped walking robot with Chebyshev-Pantograph leg mechanisms (Liang et al., 2008); a novel biologically inspired tripod walking robot (Liang et al., 2009 & 2011); a new waist-trunk system for biped humanoid robots (Carbone et al., 2009; Liang et al., 2010; Liang & Ceccarelli, 2010). In details, the content of each section are organized as follows.

In the first section, operation analysis of a Chebyshev-Pantograph leg mechanism is presented for a single DOF biped robot. The proposed leg mechanism is composed of a Chebyshev four-bar linkage and a pantograph mechanism. Kinematic equations of the proposed leg mechanism are formulated and programmed in Matlab® environment for a computer oriented simulation. Simulation results show the operation performance of the proposed leg mechanism with suitable characteristics. A parametric study has been carried out to with the aims to evaluate the operation performance as function of design parameters and to achieve an optimal design solution.

In the second section, a novel tripod walking robot is presented as inspired by tripod gaits existing in nature. The mechanical design problem is investigated by considering the peculiar requirements of leg mechanism to have a proper tripod walking gait. The proposed tripod walking robot is composed of three leg mechanisms with linkage architecture. The proposed leg mechanism is modeled for kinematic analysis and equations are formulated for simulation. A program has been developed in Matlab® environment to study the operation performance of the leg mechanism and to evaluate the feasibility of the tripod walking gaits. Simulation results show operation characteristics of the leg mechanism and feasible walking ability of the proposed tripod walking robot.

In the third section, a new torso design solution named waist-trunk system has been proposed for biped humanoid robots. The proposed waist-trunk system is composed of a six DOFs parallel manipulator and a three DOFs orientation parallel manipulator, which are connected in a serial chain architecture. In contrast to the traditional torso design solutions, the proposed new waist-trunk system has a high number of DOFs, great motion versatility, high payload capability, good stiffness, and easy-operation design features. A 3D model has been built in Matlab® environment by using its Virtual Reality (VR) toolbox. Kinematic simulations have been carried out for two operation modes, namely walking mode and manipulation mode. Operation performances have been evaluated in terms of displacements, velocities, and accelerations. Simulation results show that the simulated waist-trunk system can be very convenient designed as the torso part for humanoid robots.

2. A single DOF biped robot

A suvery of existing biped robots shows that most of their leg mechanisms are built with an anthropomorphic architecture with three actuating motors at least at the hip, knee, and ankle joints. These kinds of leg mechanisms have an anthropomorphic design, and therefore they show anthropomorphic flexible motion However, mechanical design of these kinds of leg systems is very complex and difficult. Additionally, sophisticated control algorithms and electronics hardware are needed for the motion control. Therefore, it is very difficult and costy to build properly a biped robot with such kinds of leg mechanisms.

A different methodology can be considered such as constructing a biped robot with reduced number of DOFs and compact mechanical design. At LARM, Laboratory of Robotics and Mechatronics in the University of Cassino, a research line is dedicated to low-cost easy-operation leg mechanism design. Fig. 1 shows a prototype of a single DOF biped robot fixed on a supporting test bed. It consists of two leg mechanisms with a Chebyshev-Pantograph linkage architecture.The leg mechanisms are connected to the body with simple revolute joints and they are actuated by only one DC motor through a gear box. The actuated crank angles of the two leg mechanisms are 180 degrees synchronized. Therefore, when one leg mechanism is in non-propelling phase another leg mechanism is in propelling phase and vice versa. A big U shaped foot is connected at the end of each leg mechanism with a revolute joint equipped with a torsion spring. The torsion spring makes the foot contact with the ground properly so that it has adaptability to rough terrain and the walking stability of the biped robot is improved.

Fig. 1. A prototype of a single DOF biped robot with two Chebyshev-Pantograph leg mechanisms at LARM

2.1 Mechanism description

The built prototype in Fig. 1 consists of two single DOF leg mechanisms, which is composed of a Chebyshev four-bar linkage LEDCB and a pantograph mechanism PGBHIA, as shown in Fig. 2. The Chebyshev mechanism LEDCB can generate an ovoid curve for the point B, so that the leg mechanism can perform a rear-forth and up-down motion in sagittal plane with only one actuation motor. In Fig. 2, the crank is LE, the rocker is link CD, and the coupler triangle is EDB. Joints at L, C, and P are fixed on the body of the biped robot. The offsets a, p, and h between them will greatly influence the trajectory shape of point A. The pantograph mechanism PGBHIA is used to amplify the input trajectory of point B into output trajectory with the same shape at point A. In particularly, unlike the traditional design solution, the point P is fixed on the body of the robot instead of the point B in order to have a more compact robust design. However, drawbacks will exist and in this work the aim is to maintain them within certain limits. The amplify ratio of the pantograph mechanism depends on the length of link HI and link IA or the ratio of PA and PB.

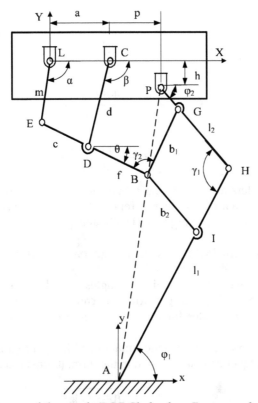

Fig. 2. A kinematic scheme of the single DOF Chebyshev-Pantograph leg mechanism

2.2 A kinematic analysis

A kinematic analysis has been carried out in order to evaluate the operation performance of the single DOF leg mechanism by using Matlab® programming. Actually, the pantograph mechanism amplifies the input motion that is produced by the Chebyshev linkage, as well as parameters p and h affect location and shape of the generated ovoid curve. A kinematic study can be carried out separately for the Chebyshev linkage and pantograph mechanism.

A scheme of the Chebyshev four-bar linkage LEDCB with design parameters is shown in Fig. 2. When the crank LE rotates around the point L an output ovoid curve can be traced by point B. Assuming a reference frame XY fixed at point L with X axis laying along in the direction of straight line LC, it is possible to formulate the coordinates of point B as a function of input crank angle α in the form, (Ottaviano et al., 2004),

$$x_B = m \cos\alpha + (c + f) \cos\theta$$

$$y_B = -m \sin\alpha - (c + f) \sin\theta \tag{1}$$

where

$$\theta = 2 \tan^{-1}\left(\frac{\sin\alpha - (\sin^2\alpha + B^2 - D^2)^{1/2}}{B + D}\right) \tag{2}$$

and

$$B = \cos\alpha - \frac{a}{m} \qquad (3)$$

$$C = \frac{a^2 + m^2 - c^2 + d^2}{2\,m\,d} - \frac{a}{d}\cos\alpha$$

$$D = \frac{a}{c}\cos\alpha - \frac{a^2 + m^2 - c^2 + d^2}{2\,m\,c}$$

The five design parameters a, m, c, d, and f characterize the Chebyshev four-bar linkage, which have a fixed ration with each other as reported in (Artobolevsky, 1979). A numerical simulation can be carried out by using Eqs. (1), (2), and (3) with proper value of the design parameters.

The pantograph mechanism PGBHIA with design parameters is shown in Fig. 2. The point P is fixed and point B is connected to the output motion that is obtained by the Chebyshev four-bar linkage. The transmission angles γ_1 and γ_2 are important parameters for mechanism efficiency. A good performance can be ensured when $|\gamma_i\text{-}90°| < 40°$ (i=1, 2) according to practice rules for linkages as reported in (Hartenberg and Denavit, 1964).

Referring to the scheme in Fig. 2, kinematic equations of the pantograph mechanism can be formulated after some algebraic manipulation in the form, (Ottaviano et al., 2004),

$$\varphi_1 = 2\tan^{-1}\frac{1 - \sqrt{1 + k_1^2 - k_2^2}}{k_1 - k_2}$$

$$\varphi_2 = 2\tan^{-1}\frac{1 - \sqrt{1 + k_2^2 - k_4^2}}{k_3 - k_4} \qquad (4)$$

with

$$k_1 = \frac{x_B - p}{y_B - h}$$

$$k_2 = \frac{b_1^2 + y_B^2 + x_B^2 - (l_2 - b_2)^2 + p^2 + h^2 - 2px_B - 2hy_B}{2b_1(y_B - h)}$$

$$k_3 = \frac{p - x_B}{y_B - h}$$

$$k_4 = \frac{-b_1^2 + y_B^2 + x_B^2 + (l_2 - b_2)^2 + p^2 + h^2 - 2px_B - 2hy_B}{2(l_2 - b_2)(y_B - h)} \qquad (5)$$

Consequently, from Fig. 2 transmission angles γ_1 and γ_2 can be evaluated as $\gamma_1 = \varphi_1 + \varphi_2$ and $\gamma_2 = \pi - \theta - \varphi_1$, respectively. The coordinates of point A can be given as

$$x_A = x_B + b_2\cos\varphi_2 - (l_1 - b_1)\cos\varphi_1$$

$$y_A = y_B - b_2\sin\varphi_2 - (l_1 - b_1)\sin\varphi_1 \qquad (6)$$

By using Eqs. (4), the transmission angles γ_1 and γ_2 can be computed to check the practical feasibility of the proposed mechanism. By derivating Eqs. (6), the motion velocities of point A can be easilty computed. Accelerations can be also computed through a futher derivative of the obbtained equations of velocities. Similarly, the velocities and accelerations at point B can be computed through the first and second derivatives of Eqs. (1), respectively.

By using velocity and acceleration analysis for the generated ovoid curve, kinematic performance of the proposed leg mechanism can be evaluated.

2.3 Simulation results

A simulation program has been developed in Matlab® environment to study kinematic performance of the proposed leg mechanism, as well as the feasible walking ability of the single DOF biped robot. The elaborated code in m files are included in the CD of this book. Design parameters of the simulated leg mechanism are listed in Table 1.

a	b	c	d	h	m
50	20	62.5	62.5	30	25
f	p	l_1	l_2	b_1	b_2
62.5	30	300	200	75	150

Table 1. Design parameters of a prototype leg mechanism at LARM with structure of Fig. 2 (sizes are in mm)

Examples of simulation results of one leg mechanism are shown in Fig. 3. When the input crank LE rotates around point L with a constant speed, the motion trajectories of point A and point B can be obtained in the form of ovoid curves. A scheme of the zoomed view of the ovoid curve in Fig. 3(a) is shown in Fig. 3(b) in which four characteristic angles of the input crank actuation are indicated. The dimension of the ovoid curve is characterized by the length L and height H. The generated ovoid curve is composed of an approximate straight-line and a curved segment with a symmetrical shape. The straight-line segment starts at the actuation angle α=90 degs and ends at α=270 degs. Actually, during this 180 degs interval the leg mechanism is in the non-propelling phase and it swings from rear to forth. During the next 180 degs interval the actuation angle goes from α=270 degs to α=90 degs corresponding to the coupler curve segment. In this period, the foot grasps the ground and the leg mechanism is in the propelling phase. The leg mechanism is in a almost stretched configuration when α=0 deg just as the leg scheme shows in Fig. 3(a).

Fig. 4 shows simulation results of the biped robot when it walks on the ground. In Fig. 4, the right leg is indicated with solid line when in contact with the ground and the crank actuation angle is at α=0 deg, the left leg is indicated with dashed line when the crank is at angle α=180 degs and it swings from rear to forth. The trajectories of points A and point B are also plotted as related to the non-propelling phase. It is noted that at the beginning and the end of the trajectory, the density of the points is higher than in the middle segment. Since the time periods are same between each plotted points, the velocity of the swinging leg mechanism in the middle is higher than that at the start and end of one step.

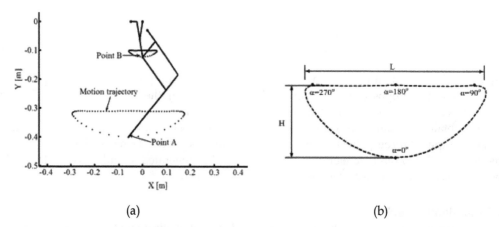

(a)   (b)

Fig. 3. Simulation results of one leg mechanism: (a) computed trajectories of points A and B; (b) generated ovoid curve at point A

A scheme of the biped motion and trajectories of critical points are shown in Fig. 5(a). Referring to Fig. 5(a), when the leg mechanism is in a non-propelling phase, it swings from rear to forth and the supporting leg propels the body forward. The swinging leg mechanism has a relative swing motion with respect to the supporting leg mechanism. Therefore, the velocity of point B1 in a non-propelling phase with respect to the global inertial frame is larger than that during a supporting phase. This is the reason why the size of curve a-b-c is larger than that the size of curve c-d-e in Fig. 5(b) even if the Chebyshev mechanisms produce the motion with only 180 degs phase differences at the points B1 and B2.

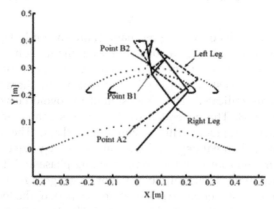

Fig. 4. Simulation results for motion trajectories of the leg mechanisms during biped walking

Fig. 5(b) shows the trajectories of points A1, A2, B1, and B2 in a biped walking gait. The trajectories are plotted with solid lines for the right leg mechanism and with dashed lines for the left leg mechanism, respectively. The motion sequences of points B1 and A1 are indicated with alphabet letters from a to e and a' to e', respectively. In Fig. 5(b), the trajectory segments a-b-c of point B1 and a'-b'-c' of point A1 are produced by the right leg mechanism while it swings from rear to forth. The trajectory segments c-d-e are produced while the right leg is in contact with the ground. Correspondingly, c', d', and e' are at the same point. The trajectories of points A2 and B2 for left leg mechanism are similar but have 180 degs differences with respect to the right leg mechanism. There are small circles in the trajectories of point B1 and point B2 during the transition of the two walking phases. This happens because there is a short period of time during which both legs are in contact with the ground and a sliding back motion occus for the body motion of the biped robot.

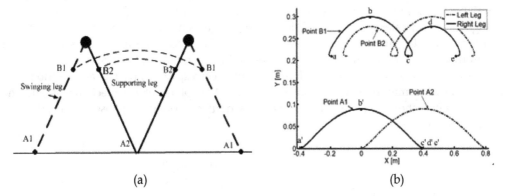

| (a) | (b) |

Fig. 5. Simulation results of biped walking: (a) motion trajectories of the leg mechanisms; (b) a characterization of the computed trajectories of points A1, A2, B1, and B2

Fig. 6(a) shows plots of the computed transmission angles γ_1 and γ_2 of the right leg mechanisms as function of the input crank angles α_1. The value of the transmission angles are computed between 50 degs and 120 degs. The transmission angles for left leg mechanism have 180 degs time differences. Therefore, the proposed leg mechanism has an efficient motion transmission capability. Fig. 6(b) shows the computed plots for angles φ_1 and φ_2 of the right leg mechanism. The value of φ_1 is between 18 degs and 100 degs as a good contact with the ground. The value of φ_2 is between -5 degs and 100 degs and there is no confliction between the legs and body.

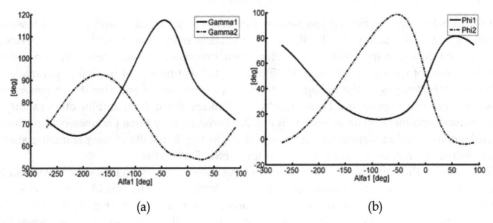

| (a) | (b) |

Fig. 6. Characterization angles of the right leg mechanism as function of angle α_1: (a) angles γ_1 and γ_2; (b) angles φ_1 and φ_2

The acceleration of point A is computed by using kinematics equations, which are computed in Matlab® m files. Fig. 7(a) shows the computed acceleration values of point A along X axis and Y axis, respectively. Similarly, Fig. 7(b) shows the accelerations of point P on the body of the biped robot.

In Fig. 7(a), the acceleration of point A at the end of leg mechanism is computed between -1 ms² to 10 m/s² along X axis and between -10.5 m/s² to -3.5 m/s² along Y axis. The acceleration along X axis reaches the maximum value when the input crank angle is at t=0.5 s (α=20 degs) and the minimum value when it is at t=7.3 s (α=325 degs).

In Fig. 7(b), the acceleration at point P is computed between -2.3 m/s² to 9 m/s² along X axis and between -10.2 m/s² to -0.2 m/s² along Y axis. The acceleration in X axis reaches the maximum value when one leg mechanism is in the middle of supporting phase and acceleration in Y axis reaches the minimum value, correspondingly. The acceleration in X axis reaches the minimum value during the transition phase of leg mechanisms and the negative value shows that the biped robot in a double supporting phase and produces a back sliding motion.

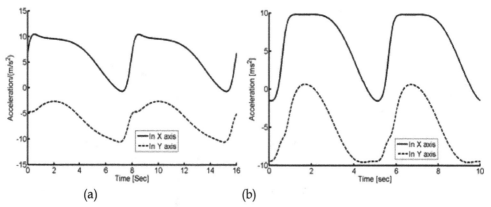

(a)   (b)

Fig. 7. Computed accelerations during one biped walking gait: (a) accelerations of point A in X and Y axes; (b) accelerations of point P in X and Y axes

An optimal design of the leg mechanism can perform an efficient and practical feasible walking gait. By using the flexibility of Matlab® environment with the elaborated simulation codes. A parametric study has been proposed to characterize the operation performance of the proposed single DOF biped robot as function of its design parameters. Actually, the lengths of the linkages determine a proper shape and size of the generated ovoid curve that is produced by the Chebyshev linkage through an amplification ration of the pantograph mechanism as shown in Fig. 2. Therefore, only three parameters a, p, and h can be considered as significant design variables. In Fig. 8, results of the parametric study are plotted as function of parameter a as output of Matlab® m files.

By increasing the value of parameter a, size of the ovoid curve is decreased in X axis and is increased in Y axis as shown in Fig. 8(a). Particularly, the ovoid curve with an approximately straight line segment is obtained when a=0.05 m. Fig. 8(b) shows the corresponding trajectories of COG (center of gravity) and the feet of swinging leg when the other leg mechanism is in contact with the ground. The step length L decreases and step height H increases as function of the value of parameter a, as shown in Fig. 8(b).

In Fig. 9, results of parametric study are plotted as function of parameter p. In Fig. 9(a), by varying parameter p the ovoid curve generated at point A has only displacements along X

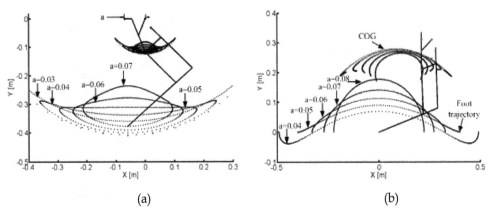

(a) (b)

Fig. 8. A parametric study of the leg mechanism as function of parameter a in Fig. 2: (a) generated ovoid curves at point A; (b) trajectories of COG and foot trajectories of the swinging leg

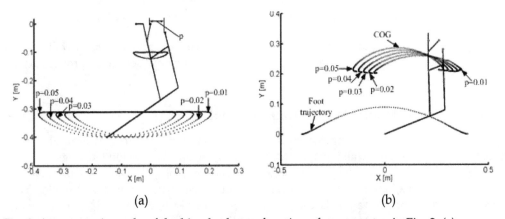

(a) (b)

Fig. 9. A parametric study of the biped robot as function of parameter p in Fig. 2: (a) generated ovoid curves at point A; (b) trajectories of COG and foot point of swinging leg

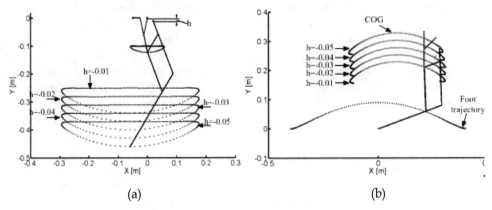

(a) (b)

Fig. 10. A parametric study of the biped robot as function of parameter h in Fig. 2: (a) generated ovoid curves at point A; (b) trajectories of COG and foot point of swinging leg

without change of the step length L and step height H. The COG of the biped robot has corresponding displacements along X axis. Similarly, by varying parameter h, the COG of the biped robot has displacements along Y axis as shown in Fig. 10.

Therefore, the position of point P determines the location of the ovoid curve without any shape change. Correspondingly, the location of COG of the biped robot can be as function of the position of point P since the mass center of the leg mechanism varies correspondingly.

The parametric study have analyzed the shape of the generated ovoid curve as function of three parameters a, p, and h. The parametric study whose main results are shown in Fig. 8, 9, and 10 has been aimed to check the motion possibility and design sensitivity of the proposed leg mechanism. Interesting outputs of the parametric study can be considered in the following aspects:

- The kinematic behaviour in terms of point trajectories is robust and well suited to walking tasks.
- Variations of main design parameters do not affect considerably main characteristics of the walking operation.
- The size of the walking step can be modified by changing the parameter a only.
- The size of height of swinging leg motion can be modified by changing the parameter h only.

Therefore, an optimized mechanical design for leg mechanism and an efficient walking gait for minimizing input crank torque can be determined by selecting proper design parameters.

3. A biologically inspired tripod walking robot

Legged locomotion in walking robots is mainly inspired by nature. For example, biped robots mimic the human walking; quadruped robots perform leg motion like dogs or horses and eight legged robots are inspired to spider-like motion (Song & Waldron, 1989; González et al., 2006). Most of animals have an even number of legs with symmetry character. With this important character animals can move easily, quickly and stably. However, among legged walking robots, biped walking robots are the human-like solutions but sophisticated control algorithms are needed to keep balance during operation (Vukobratovic, 1989). Multi legged robots have a good stable walking performance and can operate with several walking gaits. However, the number of motors increases together with legs. How to coordinate control the motors and gaits synthesis are still difficult problems.

Actually, there are some tripod walking experiences in nature, even around our daily life. A significant example of tripod walking can be recognized in old men walking with a cane. Two human legs and a walking cane as a third leg can produce a special tripod walking gait. With this kind of tripod walking gait, old people with aged or illness nervous system can walk more stably since they always keep two legs in contact with the ground at the same time. Additionally, a standing phase is more stable since there are three legs on the ground and forms a rigid triangle configuration. By talking into account of the advantages of a tripod walking gait, a novel tripod walking robot has been proposed as shown in Fig. 11.

In Fig. 11, the tripod walking robot consists of three single DOF Chebyshev-Pantograph leg mechanisms, a body frame, and a balancing mechanism, which is mounted on the top of body frame. Three leg mechanisms are installed on the body frame in a triangle arrangement with one leg mechanism ahead and two leg mechanisms rear in the same line. The main specifications of the designed model are listed in Table. 2.

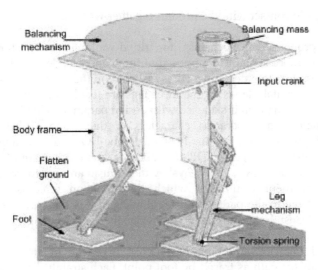

Fig. 11. A 3D model of the proposed tripod walking robot in SolidWorks® environment

Degrees of freedom	7 (3 for legs, 1 for balancing mechanism, 3 for passive ankle joints)
Weight	10 kg
Dimension	300×300×600 mm
Walking speed	0.36 km/h
Step size	300 mm/step
Walking cycle	1 sec/step

Table 2. Main specifications of the 3D model for the tripod walking robot in Fig. 11

The tripod walking robot is developed for payload transportation and manipulation purposes. The proposed design of the tripod walking robot will be capable of moving quickly with flexibility, and versatility within different environments. Therefore, in the mechanical design, particular attentions have been focused to make the tripod walking robot low-cost easy-operation, light weight, and compact. Particularly, commercial products have been extensively used in the designed model to make it easy to build. Aluminum alloy is selected as the material of the tripod walking robot since it has proper stiffness, mass density, and cheap price.

3.1 The proposed mechanical design

The mechanism design problem can be started by considering a concept of a tripod waking robot model as shown in Fig. 12(a). The scheme of the mechanism in Fig. 12(a) is a simplified structure with two DOFs that can perform a required back and forth, up and down movement in saggital plane. Actuation motors are fixed at the point C1, C2 and C3. Two feet grasp the ground at point A1 and point A2 while the third leg swings from back to forth. The two legs in contact with the ground together with the robot body form a parallel mechanism.

A scheme of the proposed leg mechanism for tripod walking robot is shown in Fig. 12(b). The tripod walking robot is mainly composed of three one-DOF leg mechanisms. The three leg mechanisms are the same design which is installed on the robot body to have a triangle configuration in horizontal plane. All the three legs are fixed on the body and actuated by DC motors. The leg mechanism is sketched with design parameters in Fig. 12(b).

The basic kinematics and operation characters of the proposed leg mechanism are investigated in the work (Liang et al., 2009). This one-DOF leg mechanism is composed of a Chebyshev four-bar linkage CLEDB and a pantograph mechanism BGMHIA. Points L, C and M are fixed on the body. The Chebyshev mechanism and pantograph mechanism are jointed together at point B through which the actuation force is transmitted from the Chebyshev linkage to the pantograph leg. Linkage LE is the crank and α is the input crank angle. The transmission angles γ_1 and γ_2 of the leg mechanism are shown in the Fig. 12(b).

When the crank LE rotates around point L, an ovoid curve with an approximate straight line segment and symmetry path as traced by foot point. Each straight line segment has a 180° phase in the crank rotation input. The straight line segment represents the supporting phase and the curve segment represents the swinging phase. When the leg mechanism operates in a supporting phase it generates a horizontal motion to points L, C and M which are fixed at the body. Therefore, the body of the robot is propelled forward without force conflict between two legs contacting the ground.

In a tripod walking gait each leg must has 2/3 period of time in supporting phase and another 1/3 time in swinging phase. In order to avoid the problem of force conflict between legs, a solution is that the two legs on the ground can produce a straight line motion in horizontal plane with the same speed and without waving in vertical direction. A careful analysis will help to define a propel operation of the leg mechanism.

A feasible solution requires that the actuation speed of the input crank is twice during swinging phase as compared with supporting phase. A_i (i=1, 2, 3) are the end points of three leg mechanisms. They trace the same ovoid curve but with 90°actuation phase differences in supporting phase. Therefore, there will be always two legs in contact with the ground and another leg swings in the air.

3.2 Simulation results
Simulations have been carried out in the Matlab® environment with suitable codes of the proposed formulation. The design parameters of the mechanisms for simulation are listed in Table.3. The rotation velocity of the input crank actuation angle is set at 270 degs/s. Each step lasts in 1/3 second for each leg, and numerical simulation has been computed for 2 seconds to evaluate a walking behavior in a stationary mode.

Chebyshev mechanism (mm)		Pantograph mechanism (mm)		Leg location (mm)
d=62.5	m=25	l_1=330	l_2=150	H_1=100
c=62.5	a=50	b_1=110	b_2=100	H_2=100
f=62.5	p=230	p=230	_	H_3=240

Table 3. Simulation parameters of the single DOF leg mechanism for the tripod walking robot

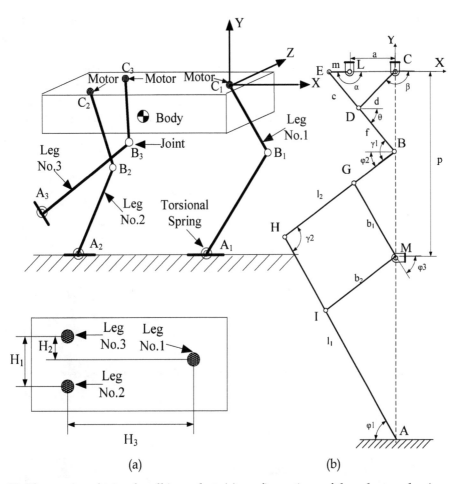

Fig. 12. The proposed tripod walking robot: (a) configurations of three leg mechanisms; (b) a scheme of one leg mechanism with design parameters

In Fig. 13, the tripod walking robot is given at initial configuration with the input crank angles α_1=180 degs, α_2=90 degs, and α_3=270 degs. At this initial time, the three legs are on the ground with two legs in supporting phase and the third leg is about to get into swinging phase.

In Fig. 14, a sequence of snapshots are shown for the tripod walking robot walks in three dimension space as computed in the numerical simulation. The trajectories of points A_i (i=1,2, 3) of the feet are depicted with small curves. In Fig. 15, the movements of the legs for tripod walking robot are shown in saggital plane. The positions of three feet are also shown in horizontal plane as referring to the computed snapshots.

As shown in Fig. 15, at each step, there are always two legs contacting the ground. Actually, a balancing mechanism can be installed on the body of the robot to adjust the gravity center between the two legs, which grasp the ground at each step. A simple rotation mechanism with a proper mass at end is likely to be installed on the body of robot as a balancing mechanism. Therefore, with a very simple control algorithm and specially sized balancing mechanism the tripod walking robot can walk with a static equilibrium even while it is walking.

A typical walking cycle for the proposed tripod walking robot can be described as following by referring to Fig. 13 and Fig. 15. The leg No.3 leaves the ground and swings from back to forth in the so-called swinging phase; at the same time the leg No.1 and the leg No.2 are in the supporting phase, since they are in contact with the ground and they propel the body forward. The speed of the input crank in leg No.3 is twice than in leg No.2 and Leg No.1. When the swinging leg No.3 touches the ground, it starts the propelling phase and the leg No.1 is ready to leave the ground. When Leg No.2 touches the ground, the tripod robot completes one cycle of walking.

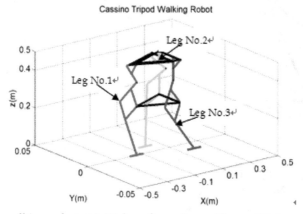

Fig. 13. The tripod walking robot at initial configuration with α_1=180 degs, α_2=90 degs, and α_3=270 degs

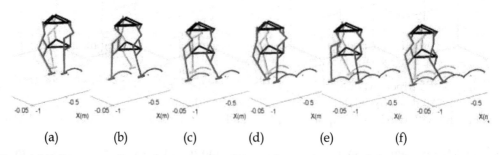

Fig. 14. Walking snapshots of the tripod walking robot as function of the input for leg motion: (a) α_1=270 degs; (b) α_1=90 degs; (c) α_1=180 degs; (d) α_1=270 degs; (e) α_1=90 degs; (f) α_1=180 degs

In order to investigate the operation characteristics and feasibility of the proposed mechanism, the plots of transmission angles γ_1, γ_2 and leg angles φ_1, φ_2 for three legs are shown as function of time in Fig. 16(a) and Fig. 16(b), respectively.

The plots are depicted for each leg. It can be found out that the transmission angle γ_1 varies between 60 degs and 170 degs and γ_2 varies between 70 degs and 120 degs. According to the kinematics rule of linkages, a feasible and effective transmission can be obtained for the proposed leg mechanism.

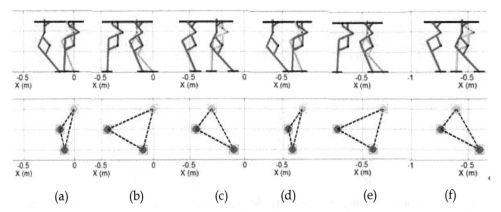

(a)   (b)   (c)   (d)   (e)   (f)

Fig. 15. Walking sequences and trajectories of the feet in saggital plane and position of the three feet in horizontal plane: (a) α_1=270 degs; (b) α_1=90 degs; (c) α_1=180 degs; (d) α_1=270 degs; (e) α_1=90 degs; (f) α_1=180 degs

The plots of leg angles φ_1, φ_2 are shown in Fig. 17. Angle φ_1 varies in a feasible region between 45 degs and 95 degs. It reaches the maximum value at the transition point from swinging phase to supporting phase and the minimum value vice versa. Angle φ_2 varies between 5 degs and 72 degs. Therefore, no conflict exists between pantograph mechanism and Chebyshev linkage in the proposed leg mechanism.

Fig. 18(a) shows plots the motion trajectories in saggital plane for points A_i (i=1, 2, 3). Dimension of the length and height for each step are depicted as L and H, respectively. These two dimension parameters are useful to evaluate walking capability and obstacles avoidance ability for the tripod walking robot. They have been computed as L=300 mm and H=48 mm for each step.

A tripod walking gait is composed of three small steps. Fig. 18(b) shows the positions of points of C_i (i=1, 2, 3) in saggital plane. It can be noted that the trajectories are approximate straight lines with very small waving. Therefore, the body of the tripod walking robot has a very small movement of less 5 mm in vertical direction and can be seem as an energy efficiency walking gait. It is computed that the body of robot is propelled forward 100 mm for each leg step. Therefore, the body is propelled forward 300 mm in a cycle of tripod walking gait. The walking speed can be computed as 0.3 m/s. However, there is a period of time that points C_2 and C_3 do not maintain the rigid body condition, but they move very slightly with respect to each other. Actually, this happens because the propelling speeds of two supporting legs are different. Therefore, a small difference of the motions between points C_2 and C_3 have been computed in the simulation of the walking gait.

Fig. 19 shows those differences between the positions of points C_i (i=1, 2, 3) as corresponding to Fig. 12(a), during the tripod walking. Fig. 19(a) shows the differences in X axis and Fig. 19(b) in Y axis, respectively. The difference in X axis is less than ΔX_2=5 mm and difference in Y axis is less than ΔY_2=1.6 mm. The difference in Y axis can be used as compliance capability during the walking also to smooth the ground contacts. The difference in X axis can be compensated by installing a passive prismatic translation joint on the leg joints at the robot body.

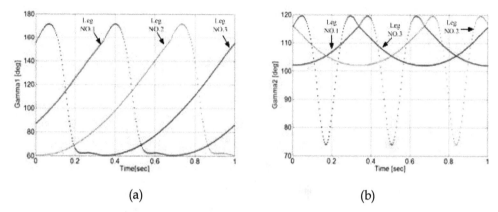

(a)   (b)

Fig. 16. The transmission angles of the three leg mechanisms during a simulated walking as function of time; (a) transmission angle γ_1; (b) transmission angle γ_2

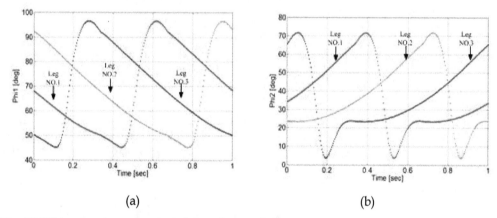

(a)   (b)

Fig. 17. The transmission angles of three leg mechanisms as function of time; (a) transmission angle φ_1; (b) transmission angle φ_2

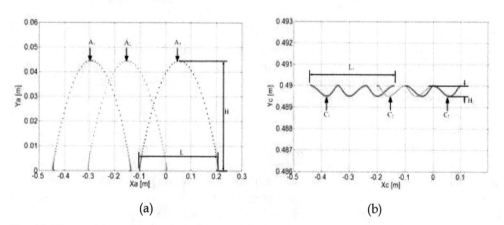

(a)   (b)

Fig. 18. The position of point A and point C in Saggital XY plane for three legs; (a) positions of points A_i (i=1, 2, 3); (b) positions of points C_i (i=1, 2, 3)

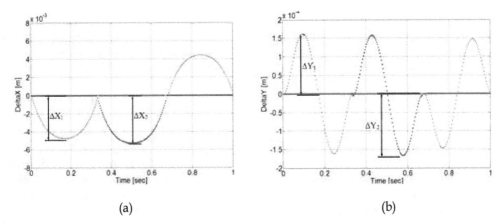

Fig. 19. The errors between points C_i (i=1, 2, 3) as function of time: (a) errors in X axis; (b) errors in Y axis

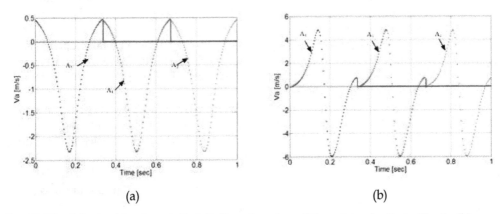

Fig. 20. The velocity of points A_i (i=1, 2, 3), as function of time: (a) velocity in X axis; (b) velocity in Y axis

The plots of velocity at points A_i (i=1, 2, 3) in X and Y axis are shown in Fig. 20(a) and .Fig. 20(b), respectively. It can be noted that the velocity reaches the maximum value when the legs move to the highest point in a swinging phase in X axis. At the same point the velocity in Y axis is zero and the sign of velocity is changed. In the supporting phase because points A_i (i=1, 2, 3) are on the ground, the velocity is zero. Since the input crank speed is twice time in swinging phase than that in supporting phase, the plots are discontinuous at the transition point. Actually, this can be modeled as an impact between feet and ground that can be smoothed by the above mentioned differences in the paths of C_i points.

Matlab® programming has been suitable and indeed efficient both for performance computation and operation simulation by using the formulated model for the design and operation of the proposed tripod walking robot.

4. A New waist-trunk system for humanoid robots

Humanoid robots are designed as directly inspired by human capabilities. These robots usually show kinematics similar to humans, as well as similar sensing and behaviour.

Therefore, they can be better accommodated in our daily life environment (home, office, and other public places) by providing services for human beings (Kemp et al., 2008). This research field has attracted large interests since two decades and a lot of prototypes have been built in the laboratories or companies. Significant examples of biped humanoid robots can be indicated for example in ASIMO developed by HONDA Corporation (Sakagami et al., 2002), HRP series developed at AIST (Kaneko et al., 2004), WABIAN series at the Takanishi laboratory in the Waseda University, Japan (Ogura et al., 2006), HUBO series built at KAIST in Korea (Ill-Woo et al., 2007), and BHR series built in the Beijing Institute of Technology, China (Qiang et al., 2005).

A survey on the current humanoid robots shows that their limbs (arms and legs) are anthropomorphically designed as articulated link mechanisms with 6 or 7 DOFs. However, torsos of humanoid robots are generally treated as rigid bodies, which are passively carried by the biped legs. The torsos of the existing humanoid robots like ASIMO, HRP, and HUBO have almost a box shaped body with a small number of DOFs. A motivation of this kind of designs is that the torso is used to store the computer, battery, sensors, and other necessary devices, so that the whole system can be designed as compact, robust, and stiff. In addition, due to mechanical design difficulties and complexity of controlling multi-body systems, torsos have been designed by using serial mechanism architectures. However, this kind of designs introduces several drawbacks, which give limitations on the motion capability and operation performances for humanoid robots (Carbone et al., 2009). Therefore, it is promising to design an advanced torso for humanoid robots by adopting parallel mechanisms with a relative high number of DOFs.

Actually, the human torso is a complex system with many DOFs, and plays an important role during human locomotion such as in walking, turning, and running. Humans unconsciously use their waists and trunks to perform successfully tasks like bending, pushing, carrying and transporting heavy objects. Therefore, an advanced torso system is needed for humanoid robots so that they can be better accommodated in our daily life environment with suitable motion capability, flexibility, better operation performances, and more anthropomorphic characteristics.

In the literature, there are few works on design and control issues of the torso system for humanoid robots. A humanoid robot named WABIAN-2R has been developed at Takanishi laboratory in the Waseda University with a 2 DOFs waist and 2 DOFs trunk, (Ogura et al., 2006). The waist and trunk of WABIAN-2R is a serial architecture and it is used for compensating the moment that is generated by the swinging legs when it walks, and to avoid the kinematics singularity in a stretched-knee, heel-contact and toe-off motion. A musculoskeletal flexible-spine humanoid robot named as Kotaro has been built at the JSK laboratory in the University of Tokyo. Kotaro has an anthropomorphic designed trunk system with several DOFs and a complicated sensor system, and it is actuated by using artificial muscle actuators. However, it is not able to walk, (Mizuuchi, 2005). A 3 DOFs parallel manipulator named as CaPaMan2 bis at LARM has been proposed as the trunk module for a low-cost easy-operation humanoid robot CALUMA with the aim to keep balance during walking and for manipulation movements, (Nava Rodriguez et al., 2005). However, these torso systems are fundamentally different from the proposed waist-trunk system.

4.1 A new waist-trunk system
Human torso is an important part of human body. It can be recognized as the portion of the human body to which the neck, upper and lower limbs are attached. Fig. 21(a) shows a

scheme of the skeleton of human torso. It can be noted that the human torso consists of three main parts: thorax, waist, and pelvis (Virginia, 1999). The rib cages and spine column of the upper part contribute to thorax. In the thorax, the heart and lungs are protected by the rib cage. The human spine is composed of 33 individual vertebrae, which are separated by fibrocartilaginous intervertebral discs and are secured to each other by interlocking processes and binding ligaments. In particular, the lumbar spine, which is the waist segment, is the most important and largest part of human spine. The main function of the lumbar spine is to bear the weight of the human body. The spine is connected with the pelvis by sacrum and the pelvis is connected with two femurs in the lower part. Additionally, there are hundred pairs of muscles, flexible tendons, and ligaments, complex blood and nervous system with different functions to make a human torso an important part of the human body. Since the human torso is composed of three portions, namely the thorax, waist, and pelvis. In Fig. 21(a), three black rectangles have been indicated on the skeleton of the human torso as reference platforms for the thorax, waist, and pelvis, respectively. Fig. 21(b) shows the corresponding positions of these three parts on the human spine. Therefore, a simply model with three rigid bodies has been proposed as shown in Fig. 22(a), which is expected to imitate the function of human torso during difference human movements.

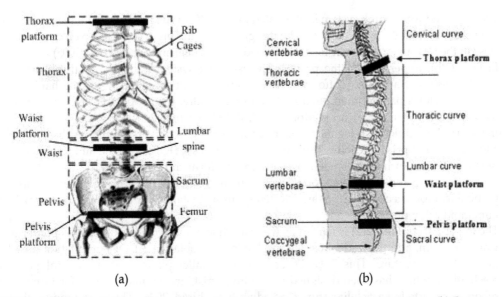

(a)   (b)

Fig. 21. Schemes of human torso with reference platforms: (a) skeleton structure; (b) S curve of a human spine

The proposed model in Fig. 22(a) is composed of three rigid bodies namely thorax platform, waist platform, and pelvis platform, from the top to the bottom, respectively. The thorax platform can be connected together to the waist platform by using parallel mechanism with suitable DOFs, which has been named as the trunk module. The thorax platform is expected to imitate the movements of human thorax. Arms, neck, and head of humanoid robots are assumed to be installed a connected to the thorax platform. The pelvis platform is connected to the waist platform with suitable mechanism, which has been named as the waist module. Two legs are expected to be connected to pelvis platform.

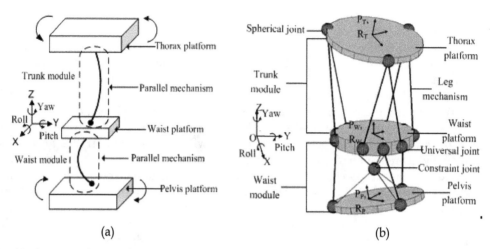

(a)   (b)

Fig. 22. A new waist-trunk system for humanoid robots: (a) a model for imitating the movements of human torso; (b) the proposed new waist-trunk system as modeled in Matlab® environment

The proposed waist-trunk system is illustrated in Fig. 22(b) as a kinematic model that has been elaborated in Matlab® environment. The design sizes of the proposed waist-trunk system are close to the human torso dimensions as reported in (Kawuchi and Mochimaru, 2005).

In Fig. 22(b), upper part of the proposed waist-trunk system is named as trunk module, which consists of a thorax platform, a waist platform, and six identical leg mechanisms to obtain a 6 DOFs parallel manipulator structure. Actually, the proposed trunk module has the same structure of a Stewart platform (Tsai, 1999; Ceccarelli, 2004).

In the trunk module, each leg mechanism is composed of a universal joint, a spherical joint, and an actuated prismatic joint. The trunk module has six DOFs with the aim to imitate the function of human lumbar spine and thorax to perform three rotations (flexion-extension, lateral-bending, and transverse-rotation movements) and three translation movements. In particular, head, neck, and dual-arm systems can be installed on the thorax platform in a humanoid robot design

The lower part in Fig. 22(b) is named as waist module, which consists of a pelvis platform, a waist platform, and three identical leg mechanisms to obtain a 3 DOFs orientation parallel manipulator structure. This 3 DOFs orientation parallel platform is a classical parallel mechanism, which has been designed as the hip, wrist, and shoulder joints for humanoid robots as reported in a rich literature (Sadjadian & Taghirad, 2006). The waist module shares waist platform with the trunk module but the leg mechanisms are installed on the counter side in a downward architecture.

The pelvis platform is connected to the waist platform with three leg mechanisms and a passive spherical joint. There are six bars connected with the passive spherical joint with the waist platform and pelvis platform with the aim to make it very stiff. The waist module is an orientation platform and has three rotation DOFs for yaw, pitch, and roll movements. The rotation center is a passive spherical joint, which plays a role like the symphysis pubis in the human pelvis to carrry the weight of the human body. The waist module is aimed to imitate the function of human pelvis during walking, running, and other movements. In particular, a biped leg system can be connected to the moving pelvis platform.

4.2 A kinematic simulation

Simulations have been carried out with the aim to evaluate the operation feasibility of the proposed waist-trunk system for a biped humanoid design solution. In Fig. 23(a) a biped humanoid robot with the proposed waist-trunk system has been modeled in Matlab® virtual reality toolbox environment by using VRML language (Vitual Reality Toolbox Users' Manual, 2007). VRML is a standard file format for representing 3D interactive vector graphics, which has been extensively used in robotic system simulation applications (Siciliano and Khatib, 2008). OpenHRP® is a simulation software package developed for performing dynamic simulation of the famous HRP series humanoid robots by using VRML language (Kanehiro et al., 2004). In a VRML file, the geometric sizes and dynamics parameters of the humanoid robot can be defined as a text-based format.

In Fig. 23(a), the modeled biped humanoid robot is composed of several balls, cuboids, and cylinders with the aim to avoid the complex mechanical design of a humanoid robot. Fig. 23(b) shows the modeling details of the waist-trunk system for a biped humanoid robot. In particular, universal joint and spherical joint are modeled by using balls. Motion constraints have been applied for each joint so that they have the proposed motion capability for a mechanical design solution. The geometry sizes and dynamics parameters of the modeled VRML model are close to the design specifications of most current humanoid robots.

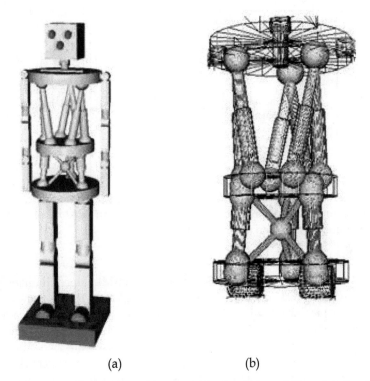

(a)   (b)

Fig. 23. 3D models in VRML: (a) a biped humanoid robot; (a) modeling details of the waist-trunk system

In Fig. 24, a simulation procedure of the biped humanoid robot in Matlab® virtual reality toolbox is shown in several steps as described in the following:

Step 1.   Movements of the moving platforms are computed according to assigned tasks. For a walking mode, motion trajectory of the waist platform is determined as based on the prescribed ZMP (zero momentum point) and COM (center of mass) trajectories. The movements of the pelvis platform are functions of the walking gait parameters. For a manipulation mode, the movements of waist platform and thorax platform depend on the locations of the manipulated objects.

Step 2.   The prescribed movements are the inputs of a motion pattern generator, where walking pattern or manipulation pattern is generated for the simulated biped humanoid robot.

Step 3.   The computed reference trajectories of the actuated joints are the inputs of a direct kinematics solver. By solving the direct kinematics of the biped humanoid robot, positions and orientations of each component can be computed.

Step 4.   The position and orientation for each component of the VRML model are updated for each step of simulation. The computed movements of the simulated biped humanoid robot are shown in animations, which are stored as videos in AVI format.

Therefore, two different operation modes of the proposed waist-trunk system can be simultated and its operation performances can be conveniently characterized by using elaborated codes included in the CD of this book.

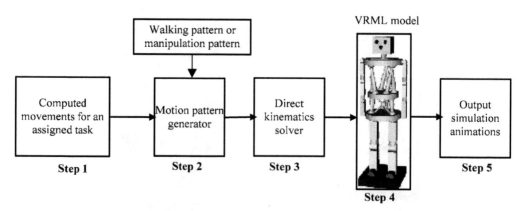

Fig. 24. A scheme for a simulation procedure of the biped humanoid robot in Matlab® virtual reality toolbox simulation environment

4.3 Simulation results

In this section, simulation results of the simulated VRML biped humanoid robot are illustrated for a walking task and a bending-manipulation task. The movements of the wais-trunk system are prescribed with suitable equations according to the assigned tasks. Operation performances of the simulated waist-trunk system have been characterized in terms of displacement, velocity, and acceleration. Simulation results show that the proposed waist-trunk system has satisfied operation characteristics as a mechanical system and has a capability of well imitating different movements of human torso.

4.3.1 Simulation of the walking mode

For a walking mode of the wais-trunk system, the waist platform is assumed to be the fixed base. Thus, positions and orientation angles of the thorax platform and the pelvis platform can be conveniently prescribed. During a normal walking, the movements of the pelvis platform and thorax platform can be prescribed by using the equations listed in Table 4.

In Table 4, $A_\varphi(v, h)$ is the magnitude of the rotation angle around roll axis as function of the walking parameters that can be determined by the walking speed v and step height l. A_θ is the magnitude of the rotation angle around pitch axis as function of the slope angle α of the ground. In particular, for a flat ground it is $A_\theta(\alpha) = 0$. $A_\psi(v, l)$ is the magnitude of the rotation angle around yaw axis as function of the walking parameters of the walking speed v and step length l. $\varphi_{W,0}$ and $\psi_{W,0}$ are the initial phase angles. $\omega = \pi/T_s$ is the walking frequency. T_s is the time period for one step of walking. The expressions in Table 4 can describe the periodical motion of the walking. A similar motion generation method is also presented in (Harada et al., 2009). The motion trajectories of the thorax platform can be prescribed similarly but in opposite motion direction in order to have a counter rotation with respect to the pelvis movements. This is aimed to preserve the angular monument generated by the lower limbs for walking stability. Particularly, only the orientation angles have been prescribed in the trunk module in the reported simulation. However, the position can be prescribed independently since the thorax platform has 6 DOFs.

	Positions (mm)	Orientation angles (degs)
Waist platform	$X_W = 0$	$\varphi_W = A_\varphi(v, h)\sin(\omega t + \varphi_{W,0})$
	$Y_W = 0$	$\theta_W = A_\theta(\alpha)$
	$Z_W = 0$	$\psi_W = A_\psi(v, l)\sin(\omega t + \psi_{W,0})$
Thorax platform	$X_T = 0$	$\varphi_T = - A_\varphi(v, h)\sin(\omega t + \varphi_{T,0})$
	$Y_T = 0$	$\theta_T = - A_\theta(\alpha)$
	$Z_T = 0$	$\psi_T = - A_\psi(v, l)\sin(\omega t + \psi_{T,0})$

Table 4. Prescribed movements for the moving platforms in a walking mode

Simulation time has been prescribed in 1.5s to simulate the function of waist-trunk system in a full cycle of humanoid robot normal walking (Ts=0.75 s/step). An operation has been simulated with 150 steps. In general, the range of motion of human pelvis is between 5 degrees and 20 degrees, and therefore, the orientation capability of the waist module has been designed within a range of 25 degrees. Thus, the waist module can imitate different movements of human pelvis through proper operations.

Fig. 25 shows the movements of the simulated biped humanoid robot, which have been simulated for two steps of walking in Matlab® environment by using the computed data in the previous analysis. The inverse kinematics analysis results have been imported to actuate the VRML model in Fig. 23. It is convenient to output the characterization values and annimations by using the flexible programming environment in Matlab®. The simulated humanoid robot shows a smooth motion which well imitates the movements of human thorax and pelvis during a walking task. In addition, it can be noted that the proposed waist-trunk system shows suitable motions to imitate the movements of the human torso during a normal walking.

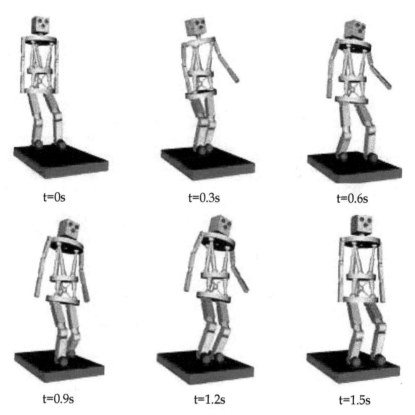

t=0s   t=0.3s   t=0.6s

t=0.9s   t=1.2s   t=1.5s

Fig. 25. Simulation snapshots of the movements of a biped humanoid robot in a walking procedure

The prescribed orientation angles of the trunk module and waist module are shown in Fig. 26(a) and Fig. 26(b). The solid and dashed lines represent rotation angle around the roll axis and yaw axes, respectively. The rotation magnitudes have been set as 20 degs and 10 degs, respectively. The dot-dashed line represents the rotation angle around the pitch axis, which has been set as a small value to avoid the computation singularity problem in the ZYZ orientation representation. The computed displacements of the prismatic joints L_i (i=1,…,6) of the trunk module are shown in Fig. 27. Fig. 28 shows the computed displacements for the prismatic joints S_k (k=1,2,3) of the waist module.
Fig. 29 and Fig. 30 show the computed velocities and accelerations for the waist and trunk modules, respectively.
It can be noted that the characterization plots are quite smooth. The proposed waist-trunk system shows a human-like behaviour for an assigned walking task because of the smooth time evaluation of the motion characteristics. The maximum velocity has been computed as 58 mm/s along Y axis for the trunk module and 120 mm/s along Y axis for the waist module. The maximum acceleration has been computed as 240 mm/s² along Y axis for the trunk module and 460 mm/s² along Y axis for the waist module. These values are feasible in proper regions for the operation of both the parallel manipulators and they properly simulate the operation of the human torso. Particularly, it can be noted that the velocity and acceleration curves of the trunk module and waist module have different signs, as an indication of the counter rotation of thorax platform and pelvis platform.

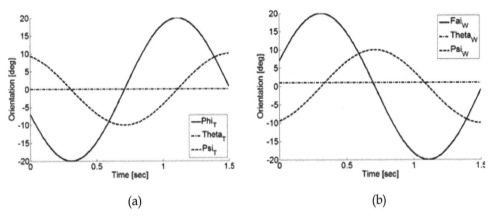

(a)   (b)

Fig. 26. Prescribed orientation angles for an operation of walking mode: (a) thorax platform; (b) pelvis platform

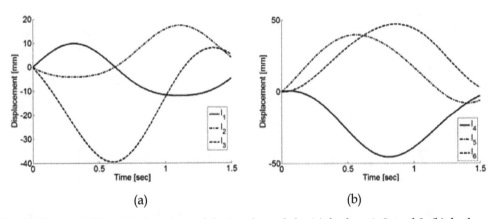

(a)   (b)

Fig. 27. Computed leg displacement of the trunk module: (a) for legs 1, 2, and 3; (b) for legs 4, 5, and 6

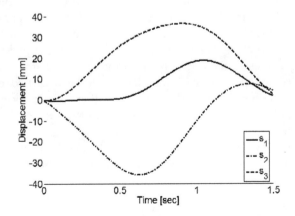

Fig. 28. Computed leg displacements of the waist module for the three legs

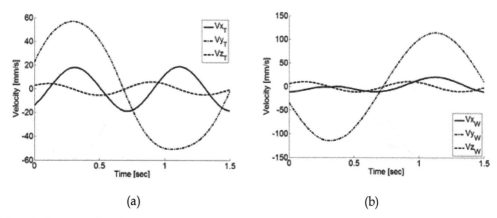

(a)   (b)

Fig. 29. Computed velocities in Cartesian space: (a) thorax platform; (b) pelvis platform

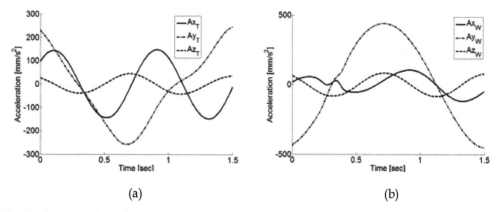

(a)   (b)

Fig. 30. Computed accelerations in Cartesian space: (a) thorax platform; (b) pelvis platform

4.3.2 Simulation of the manipulation mode

The movements of the waist module and trunk module are combined together in a manipulation mode. The waist-trunk system is a redundant serial-parallel structure with totally 9 DOFs. The inverse kinematics and motion planning are challenge issues for this peculiar serial-parallel structure. The pelvis platform has been assumed to be the fixed base, and the motion trajectories of the center point of the thorax platform and waist platform have been prescribed independently. A simulation has been carried out for a bending-manipulation procedure in order to evaluate the operation performance for a simultaneous action of the two parallel manipulator structures. The movements of the moving platforms have been prescribed by using the equations in Table 5.

Fig. 31 shows a sequence of snapshots of the simulated biped humanoid robot performing a bending-manipulation movement. The biped humanoid robot bends his torso and tries to manipulate the object that is placed on the top of a column on the ground. The double-parallel architecture gives a great manipulation capability for the biped humanoid robot, which is a hard task for current humanoid robots to accomplish. From the motion sequences in Fig. 31, it can be noted that the proposed waist-trunk system shows a suitable motion which well imitates the movements of a human torso during a bending-manipulation procedure. It is remarkable the smooth behaviour of the overall operation that makes the waist-trunk system to show a human-like motion characteristic and it can be very convenient designed as the torso part for humanoid robots.

	Positions (mm)	Orientation angles (degs)
Thorax platform	$X_T = X_{t,0} + 120 \sin(\omega t)$	$\varphi_T = 0$
	$Y_T = 0$	$\theta_T = 30 \sin(\omega t + \theta_0)$
	$Z_T = Z_{T,0} - 80 \sin(\omega t)$	$\psi_T = 0$
Waist platform	$X_W = X_W(\varphi_W, \theta_W, \psi_W)$	$\varphi_W = 0$
	$Y_W = 0$	$\theta_W = 30 \sin(\omega t + \theta_0)$
	$Z_W = Z_W(\varphi_W, \theta_W, \psi_W)$	$\psi_W = 0$

Table 5. Prescribed movements of the moving platforms for a bending-manipulation motion

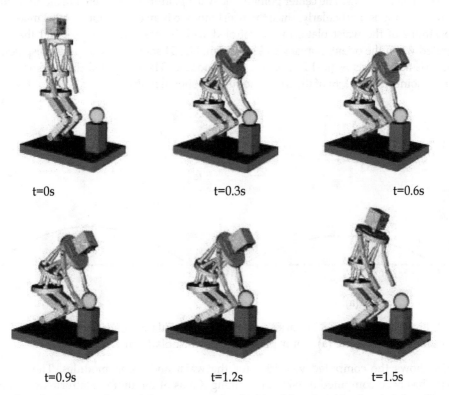

t=0s   t=0.3s   t=0.6s

t=0.9s   t=1.2s   t=1.5s

Fig. 31. Simulation snapshots of the movements of a biped humanoid robot in a bending-manipulation procedure

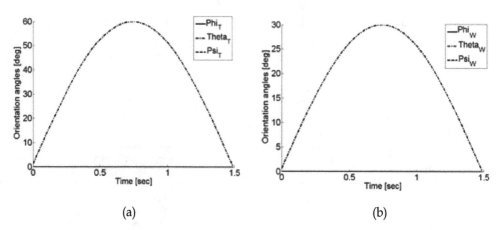

(a)   (b)

Fig. 32. Prescribed orientation angles for a simulated operation of bending-manipulation movement: (a) thorax platform; (b) waist platform

The prescribed orientation angles of the thorax platform and waist platform are shown in Fig. 32(a) and Fig. 32(b), respectively. The thorax platform rotates 60 degs around its pitch axis and the waist platform rotates 30 degs around its pitch axis. The prescribed positions are plotted in Fig. 33(a) and Fig. 33(b). The center point of the thorax platform moves 120 mm along X axis and 80 mm along Z axis. The center point of the waist platform moves 66 mm along X axis and 17 mm along Y axis. Particularly, since the waist module is an orientation parallel manipulator, the positions of the waist platform are coupled with its orientation angles and they can be computed when the orientation angles are known. Fig. 34 shows the computed displacements of the prismatic joints S_k (k=1,2,3) of the waist module. The computed displacements of the prismatic joints L_i (i=1,…,6) of the trunk module are shown in Fig. 35.

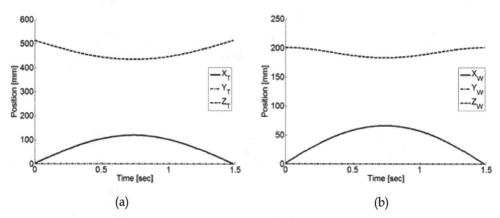

(a)   (b)

Fig. 33. Prescribed positions in Cartesian space for a simulated operation of bending-manipulation movement: (a) thorax platform; (b) waist platform

Fig. 36 shows the computed velocities for the waist and trunk modules. The maximum velocity has been computed as 600 mm/s along X axis of the thorax platform and 80 mm/s along X axis of the waist platform. Fig. 37 shows the computed accelerations in the Cartesian space with the maximum acceleration as 1000 mm/s² along X axis of the thorax platform

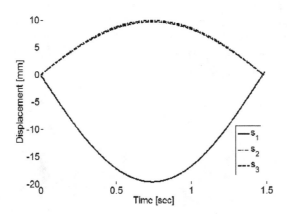

Fig. 34. Computed leg displacements of the waist module

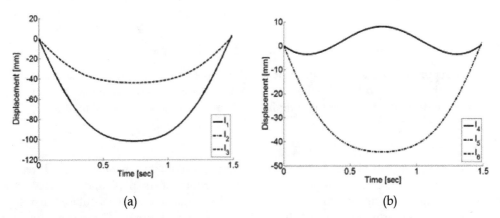

(a)   (b)

Fig. 35. Computed leg displacements of the trunk module: (a) for legs 1, 2, and 3; (b) for legs 4, 5, and 6

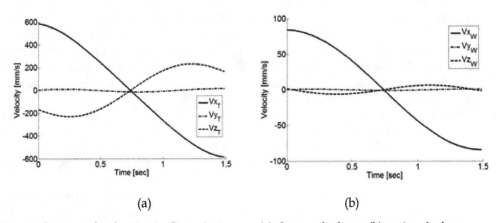

(a)   (b)

Fig. 36. Computed velocities in Cartesian space: (a) thorax platform; (b) waist platform

and 170 mm/s² along X axis of the waist platform. It can be noted that the characterization plots are quite smooth. The characterization values are feasible in proper regions for the operation of both the parallel manipulators and the proposed waist-trunk system has suitable and feasible operation performances for a robotic system as reported in the characterization plots from Fig. 32 to Fig. 37.

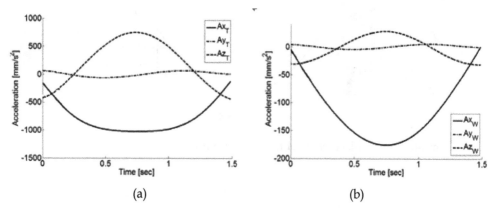

(a)   (b)

Fig. 37. Computed accelerations in Cartesian space: (a) thorax platform; (b) waist platform

From the reported simulation results, it is worth to note that a complex mechanical system such as a humanoid robot can be conveniently modeled and evaluated in Matlab® environment due to its flexible programming environment and its powerful toolbox.

5. Conclusion

In this chapter, design and simulation issues of legged walking robots have been addressed by using modeling and simulation in Matlab® environment. In particular, Matlab® is a powerful computation and simulation software package, which is quite useful for the design and operation performances evaluation of legged robotic systems. Three examples are illustrated and they have been studied for motion feasibility analysis and operation performances characterizations by taking advantages of Matlab® features. Contributions of this chapter can be indicated as follows.

A kinematic study of a Chebyshev-Pantograph leg mechanism has been carried out, and equations are formulated in the Matlab® environment. From the reported simulation results, it shows that the practical feasible operation performance of the Chebyshev-Pantograph leg mechanism in a single DOF biped robot. Additionally, a parametric study has been developed by using the elaborated Matlab® analysis code to look for an optimized mechanical design and to determine an energy efficient walking gait.

A novel biologically inspired tripod walking robot is proposed by defining suitable design and operation solution for leg mechanism. Simulation results show the proposed design performs a tripod walking gait successfully. Operation performance of the leg mechanisms and the tripod walking robot are reported and discussed by using results from Matlab® simulations.

A new waist-trunk system for humanoid robots has been proposed by using suitable parallel architectures. The proposed system shows an anthropomorphic design and operation with several DOFs, flexibility, and high payload capacity. Simulation results show that the proposed waist-trunk system can well imitate movements of human torso for walking and manipulation tasks. Additionally, the proposed design has practical feasible operation performances from the reported simulation results.

6. Acknowledgment

The first author likes to acknowledge Chinese Scholarship Council (CSC) for supporting his Ph.D. study and research at LARM in the University of Cassino, Italy for the years 2008-2010.

7. References

Song, S.M. & Waldron K.J. (1989). *Machines That Walk-The Adaptive Suspension Vehicle*, The MIT press, Cambridge, USA.

Carbone, G. & Ceccarelli, M. (2005). *Legged Robotic Systems*, Cutting Edge Robotics ARS Scientific Book, pp. 553-576, Wien, Austria.

González-de-Santos, P.; Garcia, E. & Estremera, J. (2006). *Quadrupedal Locomotion: An Introduction to the Control of Four-legged Robots*, Springer-Verlag, New York, USA.

Kajita, S. & Espiau, Bernard. (2008). *Springer Handbook of robotics, Part G-16, Legged Robots*, Springer-Verlag, Berlin Heidelberg, Germany.

Sakagami, Y.; Watanabe, R.; Aoyama, C.; Matsunaga, S.; Higaki, N. & Fujimura, K. (2002). The Intelligent ASIMO: System Overview and Integration, *Proceedings of the 2002 IEEE/RSJ International Conference on Intelligent Robots and System*, Lausanne, September 30-October 4, pp. 2478-2483.

Raibert, M. (2008). *BigDog, the Rough-Terrain Robot*, Plenary talk of the 17th IFAC world congress.

Buehler, M. (2002). Dynamic Locomotion with One, Four and Six-Legged Robots, *Journal of the Robotics Society of Japan*, Vol. 20, No.3, pp. 15-20.

Wilcox, B.H.; Litwin, T.; Biesiadecki, J.; Matthews, J.; Heverly, M.; Morrison, J.; Townsend, J.; Ahmad, N.; Sirota, A. & Cooper, B. (2007). Athlete: A Cargo Handling and Manipulation Robot for The Moon, *Journal of Field Robotics*, Vol. 24, No.5, pp. 421-434.

Matlab manual, (2007). The MathWorks, Inc. Available from http://www.mathworks.com

Liang, C.; Ceccarelli, M.; & Takeda, Y. (2008). Operation Analysis of a One-DOF Pantograph Leg Mechanism, *Proceedings of the 17th International Workshop on Robotics in Alpe-Adria-Danube Region, RAAD'2008*, Ancona, Italy, September 15-17, Paper No. 50.

Liang, C.; Ceccarelli, M.; & Carbone, G. (2009). A Novel Biologically Inspired Tripod Walking Robot, *Proceedings of the 13th WSEAS International Conference on Computers, WSEAS'2009*, Rodos Island, Greece, July 23-25, n. 60-141, pp. 83-91.

Liang, C.; Gu, H.; Carbone, G.; & Ceccarelli, M. (2010). Design and Operation of a Tripod Walking Robot via Dynamics Simulation, *Robotica*, doi: 10.1017/S026357470000000615.

Carbone, G.; Liang, C.; & Ceccarelli, M. (2009). *Using Parallel Architectures for Humanoid Robots*, Kolloquium Getriebetechnik 2009, Aachen, Germany, September 16-18, pp. 177-188.

Liang, C.; Cecarelli, M.; & Carbone, G. (2010). Design and Simulation of a Waist-Trunk System of a Humanoid Robot, *Theory and Practice of Robots and Manipulators 18th*

CISM-IFToMM Symposium on Robotics, ROMANSY'2010, Udine, Italy, July 5-8, pp. 217-224.

Liang, C.; Nava Rodriguez, N.E.. & Ceccarelli, M. (2010). Modelling and Functionality Simulation of A Waist-Trunk System with Mass Payloads, *Proceeding of the 28th Congreso Nacional de Ingeniería Mecánica, XVIII CNIM*, Ciudad Real, Spain, November 3-5, Paper no: 249.

Liang, C. & Ceccarelli, M. (2010). An Experimental Characterization of Operation of a Waist-Trunk System with Parallel Manipulator, *In Proceeding of the First IFToMM Asian Conference on Mechanism and Machine Science*, Taipei, Taiwan, October 21-25, paper no. 250042.

Artobolevsky I. (1979). *Mechanisms in Modern Engineering Design Volume V Part 1*, MIR Publishers, Moscow, Russia, pp. 405-406.

Hartenberg, R. & Denavit, J. (1964). Kinematics synthesis of linkages, McGraw-Hill Inc., New York, USA.

Ottaviano, E.; Ceccarelli M. & Tavolieri, C. (2004). Kinematic and Dynamic Analyses of A Pantograph-Leg for A Biped Walking Machine, *Proceeding of the 7th International Conference on Climbing and Walking Robots CLAWAR2004*, Madrid, Spain, September 22-24, Paper A019.

Vukobratovic, M.; Borova, B.; Surla, D.; & Storic, D. (1989). *Biped Locomotion: Dynamic Stability, Control and Application*, Springer-Verlag, New York, USA.

Kemp, C.; Fitzpatrick, P.; Hirukawa, H.; Yokoi, K.; Harada, K. & Matsumoto, Y. (2008), *Springer Handbook of Robotics, Part G. Humanoid Robots*, Springer-Verlag, Berlin Heidelberg, Germany.

Kaneko, K.; Kanehiro, F.; Kajita, S.; Hirukawa, H.; Kawasaki, T.; Hirata, M.; Akachi, K.; & Isozumi, T. (2004), Humanoid robot HRP-2, *Proceedings of 2004 IEEE International Conference on Robotics and Automation*, New Orleans, LA, USA, pp. 1083-1090.

Ogura, Y.; Shimomura, K.; Kondo, A.; Morishima, A.; Okubo, A.; Momoki, S.; Hun-ok, L. & Takanishi, A. (2006). Human-Like Walking with Knee Stretched Heel-Contact and Toe-Off Motion by A Humanoid Robot, *Proceedings of the 2006 IEEE/RSJ International Conference on Intelligence Robots and Systems*, Beijing, China, pp. 3976-3981.

Ill-Woo, P.; Jung-Yup, K.; Jungho, L. & Jun-Ho, O. (2007). Mechanical Design of the Humanoid Robot Platform, HUBO, *Advanced Robotics*, pp. 1305-1322.

Qiang, H.; Zhaoqin, P.; Weimin, Z.; Lige, Z. & Kejie, L. (2005). Design of Humanoid Complicated Dynamic Motion Based on Human Motion Capture, *Proceedings of 2005 IEEE/RSJ International Conference on Intelligent Robots and Systems*, Edmonton, pp. 3536-3541.

Mizuuchi, I. (2005). *A Musculoskeletal Flexible-Spine Humanoid Kotaro Aiming At the Future in 15 Years' Time, In Mobile Robots - Towards New Applications*, Verlag, Germany, pp. 45-56.

Nava Rodriguez, N.E.; Carbone, G. & Ceccarelli, M. (2005). CaPaMan 2bis as Trunk Module in CALUMA (CAssino Low-Cost hUMAnoid Robot), *In Proceeding of the 2nd IEEE International Conference on Robotics, Automation and Mechatronics, RAM 2006*, Bangkok, Thailand, pp. 347-352.

Virginia, C. (1999). *Bones and Muscles: An Illustrated Anatomy*, Wolf Fly Press, South Westerlo, New York, USA.

Kawauchi, M. & Mochimaru, M. (2010). AIST Human Body Properties Database, Digital Human Laboratory (AIST, Japan), Available on line: http://www.dh.aist.go.jp, 2010.

Tsai, L.-W. (1999). *Robot Analysis – The Mechanics of Serial and Parallel Manipulator*, John Wiley & Sons, New York, USA.

Ceccarelli, M. (2004). *Fundamental of Mechanics of Robotic Manipulator*, Kluwer Academic Publishers, Dordrecht, Germany.

Sadjadian, H.; & Taghirad, H.D. (2006). Kinematic, Singularity and Stiffness Analysis of the Hydraulic Shoulder: A 3-d.o.f. Redundant Parallel Manipulator, *Journal of Advanced Robotics*, Vol. 20, n. 7, pp. 763–781.

Visual Reality Toolbox User's Guide. (2007). The MathWorks, Inc. Available from http://www.mathworks.com/access/helpdesk/help/pdf_doc/vr/vr.pdf

Siciliano, B. & Khatib, O. (2008). *Springer Handbook of robotics*, Springer-Verlag, Berlin Heidelberg, Germany.

Harada, K.; Miura, K.; Morisawa, M.; Kaneko, K.; Nakaoka, S.; Kanehiro, F.; Tsuji, T. & Kajita, S. (2009). Toward Human-Like Walking Pattern Generator. *In Proceedings of the 2009 IEEE/RSJ international Conference on intelligent Robots and Systems*, St. Louis, MO, USA, pp. 1071-1077.

6

Automatic Modelling Approach for Power Electronics Converters: Code Generation (C S Function, Modelica, VHDL-AMS) and MATLAB/Simulink Simulation

Asma Merdassi, Laurent Gerbaud and Seddik Bacha
Grenoble INP/Grenoble Electrical Engineering Laboratory (G2Elab) ENSE3
Domaine Universitaire
France

1. Introduction

Modelling and simulation are useful for the analysis and design process of power electronics systems. Power electronics models of static converters are used for component sizing, for control adjustment or behaviour simulation.

In this context, average models are a good compromise between complexity, computation time and acceptable accuracy for system simulation. So, the development of averaging methods has been a topic of interest for the power electronic community for over three decades (Chetty, 1982; Middlebrook & Cuk, 1976; Sun & Grostollen, 1992).

The modelling of simple DC/DC structures is easy on the continuous conduction mode (Bass & Sun, 1998; Maksimovic et al., 2001; Maranesi & Riva, 2003; Rajagopalan, 1987; Webster & Ngo, 1992). However, the modelling complexity increases with the number of switches and the nature of the operating mode (the state sequences separated by the switch commutations). In this context, average operating analysis is becoming more and more complicated and the modelling requires skilful handling of complicated mathematical expressions which is generally time consuming.

In the literature, several automatic averaged modelling techniques have been developed however they have many limitations. As (Sun & Grostollen, 1997) that proposes a package for the modelling of PWM structures, in hard or soft switching, but the modelling is only for power electronics structures alone. The automatic modelling of a three-phase electrical machine with static converter is not possible. So, the environment of the static converter is limited, e.g. the load can not be a three-phase synchronous machine.

In the Electrical Engineering Laboratory (G2Elab), (Verdiere et al., 2003) proposes a new software architecture for the average modelling of power electronics structures, but this software is limited to the treatment of DC/DC converters in the continuous conduction mode, the classic average model is only obtained automatically.

The same problem as (Sun & Grostollen, 1997), the automatic coupling of converter with electrical machine is not considered. Finally, the computing technologies used to the topological analyze of electrical circuit (e.g. Macsyma) are not up to date.

The originality of this chapter consists to improve, on one hand, the work of (Sun & Grostollen, 1997) and (Verdiere et al., 2003), by proposing a new automatic methodology deals with several models (average and exact models) for DC/DC, DC/AC and AC/AC converters, and on other hand, by modelling converters with or without their environment (e.g. electrical machines). The particularities of this approach are the generation of models in different forms: C S-Function, VHDL-AMS, Modelica and the automatic treatment of the static converters in the continuous and discontinuous conduction mode. In this chapter, only results for the continuous conduction mode are presented.

In the first part of this chapter, the modelling approach is explained and the main steps of this modelling are detailed. A causal approach for the modelling of static converters of a power electronics applications independently of their environment (i.e its sources and loads) is also presented.

In the second part, the implementation in the software tool (AMG for Average Model Generator) dedicated to this automatic building is described. The code generation in C S-function, VHDL-AMS and Modelica is illustrated for a boost converter.

In the last part, several applications results are presented and implemented in several software (e.g Matlab/Simulink), among them a resonant converter, a three phase inverter coupled with a three-phase machine and a multi-level converter.

2. Modelling approach

The modelling approach in the software tool AMG is defined by using an a priori on the static converter behaviour and considering some hypothesis (ideal switches OFF "open circuit" and ON "short circuit", linear and invariant passive components and perfect sources).

Every model is entirely made in an automatic way from a topological description of the static converter. This description is carried out from a Netlist file extracted by using simulation software (i.e PSpice) and a dedicated simplified component library is used.

The user must define the commutation mode (i.e the cyclic sequence of switched configurations in an operating period) and the switch control sequence of the static converter to be modelled, mainly the control links of switches. The switch control and the operating mode are deduced from the analysis of the studied static converter, e.g., by using simulation software (i.e Portunus, Psim, PSpice, Simplorer, Saber, etc.). These data information given by the user are introduced into the software tool AMG.

The approach of the automatic building includes three important steps:
1. the analysis of the circuit,
2. the extraction of the state matrixes for each configuration of the static converter,
3. the building of the global state model.

Steps of modelling are presented in Fig. 1.

2.1 Topological analysis

An electrical circuit can be represented by a directed graph. The edges of this graph then connect the nodes of the circuit pairs. It is therefore possible to extract a tree of this graph corresponding to a subset of edges covering all nodes in the graph, but not forming mesh. The topological analysis is to automatically determine a tree from a simple description of the topology of the circuit. This method has been tested and used in different works (Delaye et al., 2004).

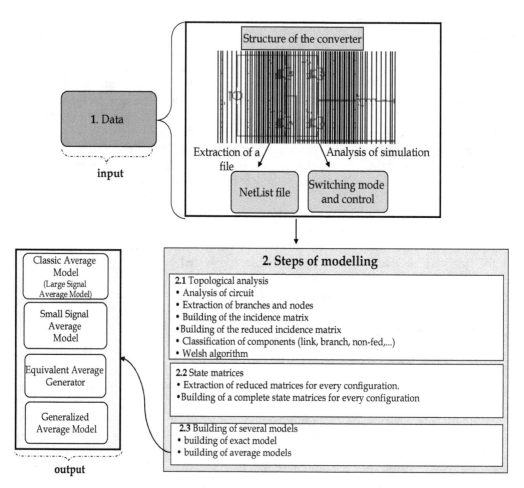

Fig. 1. Steps of modelling

Thus knowing the topology of any electrical circuit, it is possible to express the incidence matrix nodes/branches and then calculate the reduced incidence matrix.

In the incidence matrix, the rows correspond to nodes and the columns to the branches (see Fig. 2).

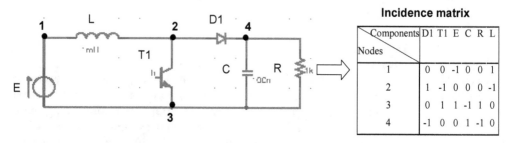

Fig. 2. Incidence matrix of a boost converter

After, this matrix is reduced for each state, according to the conduction state of switches. A switch at state ON is considered as "short-circuit": the lines corresponding to its nodes are added or subtracted (to keep a resulting line with only 0, 1 and -1). A switch at state OFF is considered as an "open circuit": the switch column is suppressed.

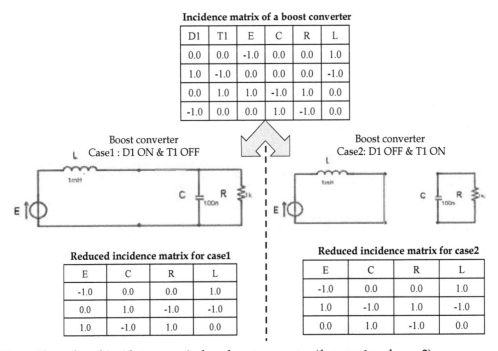

Incidence matrix of a boost converter

D1	T1	E	C	R	L
0.0	0.0	-1.0	0.0	0.0	1.0
1.0	-1.0	0.0	0.0	0.0	-1.0
0.0	1.0	1.0	-1.0	1.0	0.0
-1.0	0.0	0.0	1.0	-1.0	0.0

Boost converter
Case1 : D1 ON & T1 OFF

Boost converter
Case2: D1 OFF & T1 ON

Reduced incidence matrix for case1

E	C	R	L
-1.0	0.0	0.0	1.0
0.0	1.0	-1.0	-1.0
1.0	-1.0	1.0	0.0

Reduced incidence matrix for case2

E	C	R	L
-1.0	0.0	0.0	1.0
1.0	-1.0	1.0	-1.0
0.0	1.0	-1.0	0.0

Fig. 3. The reduced incidence matrix for a boost converter (for case 1 and case 2)

Welsh algorithm is used to extract the relations between the voltages U and currents I of the circuit (Rajagopalan, 1987). It separates links components and branches variables:
- Vmc, Vmr, Vml, Vmj are respectively the vectors of links voltages for the capacitors, the resistors, the inductors and the sources.
- Vbc, Vbr, Vbl, Ve are respectively the vectors of branches voltages for the capacitors, the resistors, the inductors and the sources.
- Ibe, Ibc, Ibr, Ibl are respectively the vectors of branches currents for the sources, the capacitors, the resistors and the inductors.
- Imc, Imr, Iml, Ij are respectively the vectors of links currents for the capacitors, the resistors, the inductors and the sources.

So, Kirchhoff laws and relations between, on the one hand, the branches voltage and current and on the other hand, between links voltage and current are easily deduced.

We propose also a new extension of the Welsh algorithm that offers the possibility to deduce:
- non fed components
- short-circuited components

Finally, the topological analysis offers a first set of circuit equations in the form of relations between on the one hand, the branch current (Ib) and the link current (Im) and on the other hand, the branch voltage (Vb) and the link voltages (Vm).

2.2 State matrices

The state matrices of the static converter are extracted for each configuration from their simplified nodes equations. By combining the equations of voltage-current relations, a reduced state system is obtained (Kuo-Peng et al., 1997).

$$\frac{d}{dt}\begin{bmatrix} V_{bc} \\ I_{ml} \end{bmatrix} = A \cdot \begin{bmatrix} V_{bc} \\ I_{ml} \end{bmatrix} + B \cdot \begin{bmatrix} V_e \\ I_j \end{bmatrix} \tag{1}$$

$$\begin{bmatrix} I_e \\ V_{mj} \end{bmatrix} = C \cdot \begin{bmatrix} V_{bc} \\ I_{ml} \end{bmatrix} + D \cdot \begin{bmatrix} V_e \\ I_j \end{bmatrix} \tag{2}$$

The equations (1) and (2) are written as (3) and (4):

$$\frac{dx}{dt} = A \cdot x + B \cdot u \tag{3}$$

$$Y = C \cdot x + D \cdot u \tag{4}$$

Where:
x : the state vector (size[n])

$$x = \begin{bmatrix} V_{bc} \\ I_{ml} \end{bmatrix} \tag{5}$$

u : the input vector (the external sources) (size[p]).

$$u = \begin{bmatrix} V_e \\ I_j \end{bmatrix} \tag{6}$$

$Y(t)$: the output vector (size[q])

$$Y = \begin{bmatrix} I_e \\ V_{mj} \end{bmatrix} \tag{7}$$

A : the state matrix (size [n,n]).
B : the input matrix (size[n,p]).
C : the output matrix (size[q,n]).
D : the feedforward matrix (size[q,p]).

2.3 Building of the several models

The global global state model of the converter is made from the state equations for each configuration of the static converter. It consists in extracting the state system for each possible topology of circuit in order to create the converter model. The approach focuses on exact and average models (classic average model, equivalent average model, generalized average model and small signal average model). The models are generated according to the nature of the power conversion (DC/DC, AC/DC) and the operating mode (continuous conduction or discontinuous conduction).

A Comparison between several models generated by AMG is presented in table 1 (Merdassi et al., 2010).

Models	Domain of validity	Results	Limitations
Exact model (Etxeberria-Otadui et al., 2002)	- DC and AC	- Most faithful model	- Not adapted for continuous control - Analyze of modes
Classic average model (Middlebrook & Cuk, 1976)	- DC-DC -Continuous conduction	- Sliding average - Not linear models	-Discontinuous conduction - Alternatives variables
Small signal average model (Bacha et al., 1994; Kanaan et al., 2005)	- DC and AC	- Linear models - Extraction of average values	-Validity around the balance operating point
Equivalent average generator (Sun & Grostollen, 1997)	-Discontinuous conduction - AC and DC	- Model of reduced dimension	- Precision missed in continuous conduction
Generalized average model (Sanders et al., 1990 ; Petitclair et al.,1996)	- Alternatives variables - DC and AC converters - Continuous conduction	- Transient mode (magnitude and phase)	- Complicated computation for higher harmonic range

Table 1. Comparison between the models generated by AMG

2.3.1 The exact model

The exact model represents the starting point to any average modelling operation (Bacha et al., 1994). Regarding the state of the various switches, the converter can be considered as a variable structure system with N possible configurations during a given commutation period T.

The state equation of each configuration of index i ($i = 1,..,N$) is defined by equation (8) in a corresponding time interval $t \in [t_{i-1}, t_i]$.

$$\frac{dx}{dt} = A_i' \cdot x + B_i' \cdot e \tag{8}$$

Where:

$$\sum_{i=1}^{N} t_i - t_{i-1} = T \tag{9}$$

T : the switching period.

N: the number of topologies (configurations) that happen during the switching period T.

x : the state vector (size [n]).

t_i : the commutation instant between configuration i and configuration $i+1$

A'_i : the state matrix for the ith configuration (size [n,n]).

B'_i : the input matrix (size [n,p]).

e : the external sources (size [p]).

By assembling these N equations, considering the continuous character of the state variable (vector x) and by assigning a discrete switching function $u_i(t) \in \{0,1\}$ or $\{-1,1\}$, representing both the operating mode and the control, the global behaviour of the static converter is formulated as the bilinear equation (10) (Sun & Grostollen, 1992).

Note that p switching functions can describe 2^p configurations.

$$\frac{dx}{dt} = A \cdot x + \sum_{i=1}^{p} u_i \cdot \left(B_i \cdot x + b_i \right) + d \tag{10}$$

Where:

x : the state vector (size [n]).

A: the global state matrix (size [n,n]).

B_i : input matrix (size [n,n])

u_i : the vector control (size [n,1]).

b_i, d: other input matrixes (size [n,1]).

2.3.2 The average model

The average value of the kth harmonic is defined by equation (11).

$$\langle x(t) \rangle_k = \frac{1}{T} \int_{t}^{t+T} x(\tau) \cdot e^{-jk\omega\tau} d\tau \tag{11}$$

Where:

$$\begin{cases} \omega = \dfrac{2\pi}{T} \\ x(\tau) = 0 \quad pour \quad \tau < 0 \end{cases}$$

$\langle x(t) \rangle_k$ represents the harmonic coefficient of range k in the decomposition of the complex Fourier series. It's average is made on a sliding window and not on the static interval.

AMG deals with the following types of static converters: DC/DC, DC/AC and AC/AC.

It allows the automatic average modelling. In the literature, the following average models (classic average model, equivalent average generator, generalized average model and small-signal average model) are well known and presented in the papers (Chen & Sun, 2005; Etxeberria-Otadui et al., 2002; Verdiere et al., 2003 & Sanders et al., 1990).

3. Causality of the static converter

For the models of loads or sources that can not be simply described by a circuit representation, a Netlist or graphical approach is limited. So, it is useful to carry out models of static converters without the description of their environment, by using only knowledge on the connection points of the converter with its environment. Then, the obtained models of the static converters are introduced into the global application.

Two possibilities are proposed: the static converter model is either oriented or non oriented, i.e. causal or non causal (Allain et al., 2009).

3.1 Oriented model

The global model of the static converter is described in a block (e.g. a C S-Function of Matlab/Simulink). Then, it is connected with the other blocks of the system. In this approach, the static converter is considered as a causal model, i.e. the solving sequence of its equations is imposed. So, the model has inputs and outputs.

The figure 4 illustrates this causality in the meaning of block representation and Bond Graph representation for a single-phase voltage inverter.

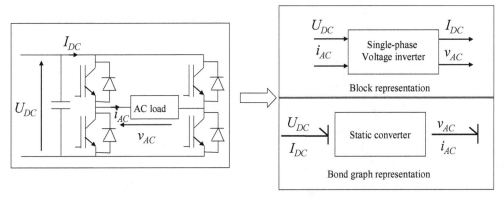

Fig. 4. Example of causality aspect for a single-phase voltage inverter

3.2 Non oriented model

The causality of the static converter can be determined by respecting the definition given in Bond Graph representation (Karnopp et al., 1990), but applied in the electrical domain. The model can be described in a non causal modelling language, e.g. VHDL-AMS, Verilog or Modelica (for the analog part). This means that the model equations are not oriented. Then the model of the static converter is connected to the models of the components thanks to the modelling language. Finally, the compilers of such languages manage automatically the causality (so the solving sequence of the model equations) at the start of the simulation running.

Practically, the static converter environment is defined with equivalent sources. Thus, a voltage source or a current source is imposed at every connection point (Mohan et al., 1989).

However, in power electronics, according to their value, inductors may be compared to current sources whereas capacitors are compared to voltages sources.

In the viewpoint of ordinary differential equation solving, the integral causality requires a numerical integration to get the state variables, i.e. the currents across the inductor and the voltages through the capacitors. So, for the external connection of the static converter, current sources (respectively voltage sources) can be used to represent inductors or inductive phenomena (respectively capacitors or capacitive phenomena).

By considering the integral causality, defined as the preferential causality in Bond Graph, the causality of capacitors and inductors are defined (see Fig.5).

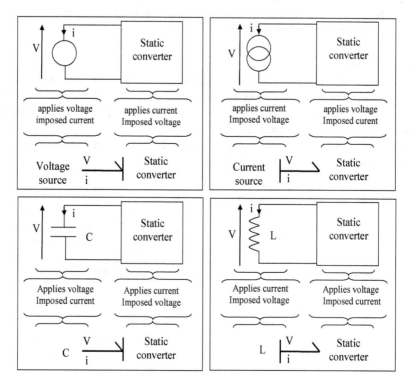

Fig. 5. Causality of a static converter

This approach proposes to generate automatically its models (exact and/or average) according to the nature of the state variables. However, to associate the generated model with other models, thanks to the static converter environment characterisation, the generated model has to give an expression for every complementary energetic variable of the equivalent sources that are connected to the converter. In this way, the current across voltage sources and the voltage through current sources have to be formulated.

The following outputs are formulated:

I_e : the sub-vector of the current in the voltage sources at the interface of the static converter.

V_{mj} : the sub-vector of the voltage on the current sources at the interface of the static converter.

Finally, models are obtained by using equation (2). This representation form of state equation is especially interesting for coupling models.

4. Implementation

AMG is implemented by using Java and Maple programming. Only symbolic treatments are in Maple. The building of the state matrixes A_i, B_i, C_i and D_i for every configuration (indexed i) is developed in Java by equations (1) and (2). The building of the exact and the average models (equations (10) and (11)) by a symbolic approach is made in Maple (Merdassi et al., 2008).

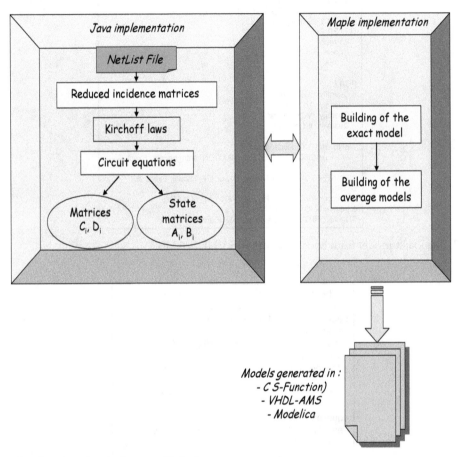

Fig. 6. Implementation in Java and Maple

In this section, an example of the generated C-S function, Modelica and VHDL-AMS code are presented for a boost converter.

```
// pour chaque entree
    InputRealPtrsType uPtrs0 = ssGetInputPortRealSignalPtrs(S,0);
    real_T          himoy = *uPtrs0[0];
    real_T          U1    = *uPtrs0[1];


    VC1moy=x[0];
    IL1moy=x[1];


dVC1moy_dt= -(VC1moy - IL1moy * R1 - IL1moy * R1 * himoy) / C1 / R1;
dIL1moy_dt= (-VC1moy + VC1moy * himoy + U1) / L1;
    dx[0] = dVC1moy_dt;
    dx[1] = dIL1moy_dt;
```

Fig. 7. The average model for a boost converter in C S-Function

```
---------- VHDLAMS MODEL hacpar_Exact ----------
LIBRARY ieee;
USE          ieee.ALL;
use ieee.math_real.all;
---------- ENTITY DECLARATION  hacpar_Exact ----------
  ENTITY hacpar_Exact  IS
GENERIC(
  R1:real;
  L1:real;
  C1:real
);
PORT  (
  Quantity  h1:in real;
  Quantity  U1:in real;
  Quantity  VC1:out real;
  Quantity  iL1:out real
);
  END ENTITY  hacpar_Exact ;
---------- ARCHITECTURE DECLARATION  arch_hacpar_Exact -------
---
ARCHITECTURE  arch_hacpar_Exact  OF  hacpar_Exact  IS
BEGIN
VC1'dot== -(VC1 - iL1 * R1 + iL1 * R1 * h1) / C1 / R1;
iL1'dot== (-VC1 + VC1 * h1 + U1) / L1;
END ARCHITECTURE arch_hacpar_Exact ;
```

Fig. 8. The exact model for a boost converter in VHDL-AMS

```
model   hacpar_Exact
external connector InputReal = input Real;
parameter Real R1(fixed=false)=0.5 ;
parameter Real L1(fixed=false)=0.5 ;
parameter Real C1(fixed=false)=0.5 ;
  InputReal h1;
  InputReal U1;
  Real VC1;
  Real iL1;

equation
der(VC1)= -(VC1 - iL1 * R1 + iL1 * R1 * h1) / C1 / R1;
der(iL1)= (-VC1 + VC1 * h1 + U1) / L1;

end hacpar_Exact ;
```

Fig. 9. The exact model for a boost converter in Modelica

5. Applications in Matlab/Simulink

Many examples of converters are implemented in Matlab/Simulink. The results are presented for several applications on the continuous conduction mode among them a resonant converter, a three phase inverter coupled with a three-phase machine (modelled by a three-phase resistor-inductor load). Finally, a multi-level converter is also taken as an example to show the interest of using AMG for big power electronic structure.

5.1 Application 1: A resonant converter

The static converter has been modelled by AMG. The studied structure has an alternative level so the generalized average model is generated for the alternative and the continuous state variables. The equivalent average generator is also tested for this application.

5.1.1 Definition of the control

The mode and the control are deducted from a classic simulation of the structure to be studied in simulation software (e.g Portunus). In the example, these signals will be known as *h1* that controls (MOS1, MOS8) and *h2* that controls (MOS3, MOS6). Their complementary $\bar{h}_1$ and $\bar{h}_2$ respectively control (MOS7, MOS2) and (MOS5, MOS4).

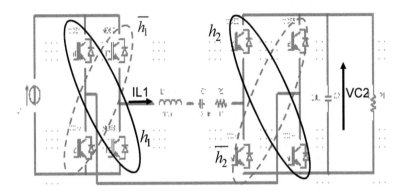

Fig. 10. Control associated to the switching cells

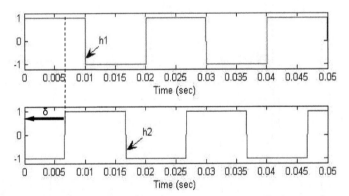

Fig. 11. Full wave control: *h1* et *h2* between [-1,1]

Square control signals are used for the switch control *h1* and *h2*. These signals are periodical and characterized by a 0.5 duty cycle and the phase shifting (δ) between the control *h1* and *h2*.

5.1.2 Simulation in Matlab/Simulink

The generalized average model allows to get the amplitude and the phase of alternative state variables. They are deduced from the real and imaginary expression of IL1.

In Fig. 13 and Fig. 14, the amplitude of the current inductor IL1 is shown for several values of delta (δ).

Finally, we remark that the envelope of the current inductor IL1 illustrates the good concordance between the model generated automatically and the real model.

For the continuous state variable (the voltage VC2), the classical average model, the exact model and the equivalent average generator are generated as shown on Fig. 15.

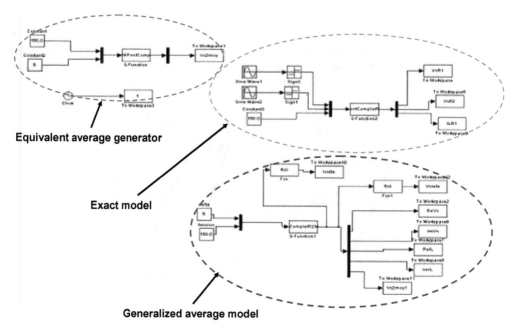

Equivalent average generator

Exact model

Generalized average model

Fig. 12. Implementation of the models for a resonant converter in Matlab/Simulink

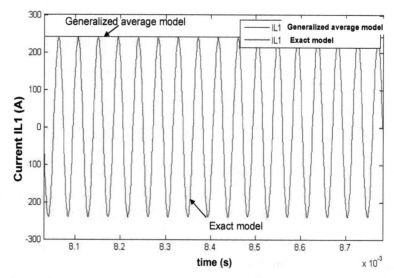

Fig. 13. Extraction of the current amplitude of IL1 from the generalized average model (for $\delta = 5$)

We can notice that, on the one hand, the average equivalent generator is not accurate for this example because it is not adapted to the converter operating in the continuous conduction mode. In fact, the equivalent average generator eliminates some dynamics of the system especially in the discontinuous conduction mode (cf. table 1). On the other hand, the classic average model is better suited for the converter.

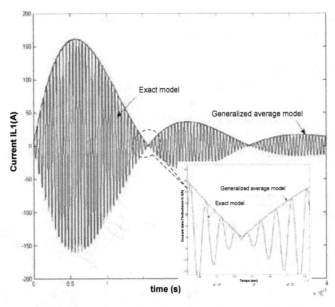

Fig. 14. Extraction of the current amplitude of IL1 from the generalized average model (for δ = 0)

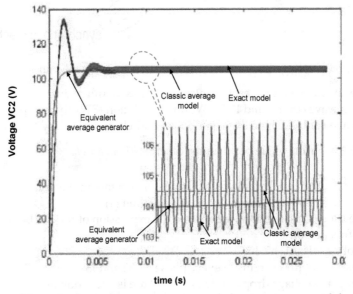

Fig. 15. Exact model, equivalent average generator and classic average model of the voltage VC2

5.2 Application 2: A three-phase voltage inverter

In this section, a voltage inverter feeding an electrical machine is studied. The proposed structure operates in continuous conduction mode, i.e., every switch commutation is controlled.

As explained above, AMG needs basic assumptions on the external equivalent electrical sources to define its environment to connect to it.

5.2.1 Definition of the control

In this static converter, each commutation cell is an inverter legs. For, one leg when a switch is controlled to be ON the other one is OFF. As a consequence, the inverter needs only three control signals corresponding to the upper switch of the switching cell (MOS1, MOS3 and MOS5). These signals will be known as $h1$, $h2$ and $h3$, respectively controls MOS1, MOS3 and MOS5. So, their complementary $\bar{h}_1, \bar{h}_2$ and $\bar{h}_3$ respectively controls MOS2, MOS4 and MOS6.

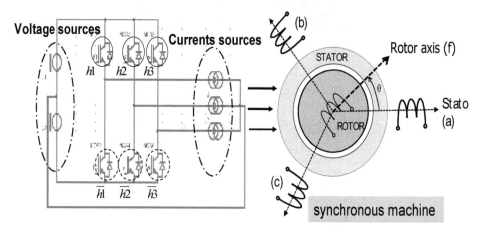

Fig. 16. Three phase voltage inverter coupled with synchronous machine

Here, for the average model, each control signal is a full wave control one. For this application, full wave control and Pulse Width Modulation (PWM) are applied.

For the full wave control, signals are periodical and characterized by a 0.5 duty cycle and $\delta = \dfrac{2\pi}{3}$ the phase shifting angle between the controls $h1$, $h2$ and $h3$.

A C S-function is obtained. It contains equations of currents in the voltage sources and the voltage on the current sources according to the equation (4).

Analyzing the code (C S-Function), we can see the expression of the DC bus current (i.e. IU1 and IU2) as a function of:
- the control signals: h1, h2, h3
- the expressions of the currents in every phase: I1, I2 and I3

In the same way, the voltage drop applied in the phase is a function of:
- the DC bus voltages: U1, U2
- the control signals: h1, h2, h3.

5.2.2 Simulation in Matlab/Simulink

The model has been mixed with the equations of a three phase RL circuit (see Fig. 18) and three-phase synchronous machine which model is a Park Model. Park's transforms and inverse Park's transforms have been added in the machine model.

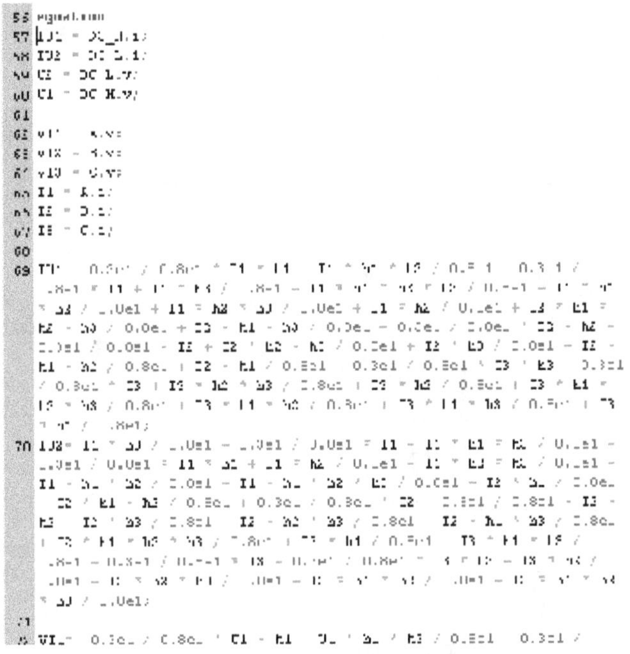

Fig. 17. Exact model for a three phase voltage inverter (in Matlab/Simulink C S-Function)

The interest of the AMG tool becomes obvious by looking at the equation generated for this very basic power converter structure. One can easily imagine the difficulty to build such a model without modelling tool, for more complex structures, such a multi-level converter.

5.3 Application 3: A multi-level converter
A multi-level converter is taken as an example to show the interest of using AMG for big power electronic structure.

5.3.1 Definition of the control
For this application, the static converter legs are supposed perfectly balanced. In this example, only the model of one leg is generated by AMG.
The switch commutations are controlled by the signals: *h1*, *h2*, *h3* and *h4* and their complementary values.
Each control signal is a pulse wave modulation. Each switching cell is control by the same sinusoidal modulation but with a $\pi/2$ phase displacement of its triangular waveform compared to the one of the switching cell where it is imbricated.

5.3.2 Simulation in Matlab/Simulink
The exact model of the current IL1 generated in C S-function (e.g Matlab/Simulink) is compared to the classical simulation from electrical software (e.g Portunus).
By using the exact model generated by AMG, simulation results represent the perfect behaviour of the multi-level converter. The same results are given with a comparison between the current IL1 obtained automatically (from the exact model by AMG) and the one obtained from the classical simulation (e.g Portunus). Finally, the automatic modelling approach is also approved for a complicated structure.

Equations generated by AMG

Vmjl1== -0.3e1 / 0.8e1 * U1 * h1 + U1 * h1 * h3 / 0.8e1 - 0.3e1 / 0.8e1 * U1 + U1 * h3 /
0.8e1 + U1 * h1 * h2 * h3 / 0.8e1 + U1 * h1 * h2 / 0.8e1 + U1 * h2 * h3 / 0.8e1 + U1 * h2 /
0.8e1 + U2 * h3 / 0.8e1 + 0.3e1 / 0.8e1 * U2 - U2 * h1 * h3 / 0.8e1 - 0.3e1 / 0.8e1 * U2 *
h1 + U2 * h2 / 0.8e1 - U2 * h2 * h3 / 0.8e1 - U2 * h1 * h2 / 0.8e1 + U2 * h1 * h2 * h3 /
0.8e1;
Vmj2== U1 * h1 * h2 * h3 / 0.8e1 + U1 * h1 * h3 / 0.8e1 - 0.3e1 / 0.8e1 * U1 * h2 - 0.3e1 /
0.8e1 * U1 + U1 * h2 * h3 / 0.8e1 + U1 * h3 / 0.8e1 + U1 * h1 * h2 / 0.8e1 + U1 * h1 / 0.8e1
+ U2 * h1 / 0.8e1 - U2 * h1 * h2 / 0.8e1 + 0.3e1 / 0.8e1 * U2 - 0.3e1 / 0.8e1 * U2 * h2 + U2
* h3 / 0.8e1 - U2 * h2 * h3 / 0.8e1 - U2 * h1 * h3 / 0.8e1 + U2 * h1 * h2 * h3 / 0.8e1;
....

/* Equations of RL added*/

dI1_dt= (double) (Vmj1-(R4*I1)) / L1;
dI2_dt = (double)(Vmj2-(R5*I2))/ L2;
dI3_dt = (double) (Vmj3-(R6*I3)) / L3;

Implementation in S-Function

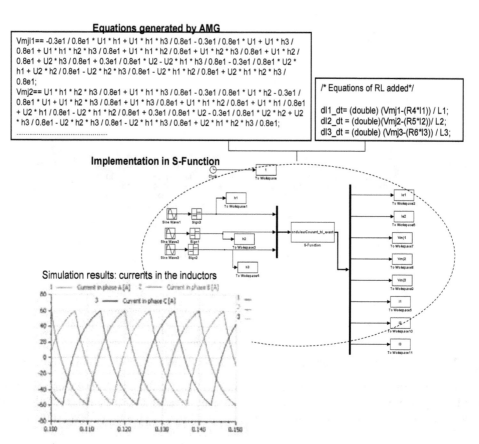

Simulation results: currents in the inductors

Fig. 18. Implementation in Matlab/Simulink: 3-phase RL load

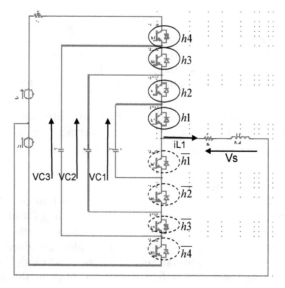

Fig. 19. One leg of the three phase multi-level circuit converter

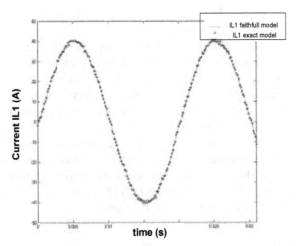

Fig. 20. Comparison between the current IL1 from the exact model (from i.e Matlab/Simulink) and the classical reference model (from i.e Portunus)

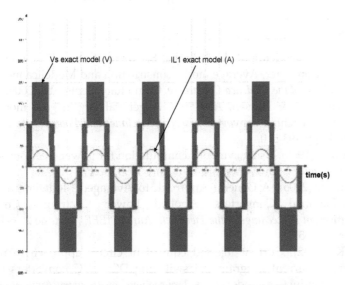

Fig. 21. Results of simulation of the exact model generated by AMG

6. Conclusion

This chapter proposes an automatic modelling process that allows to obtain different models of static converter according to the nature of the power conversion and the state variables. Every model is entirely made in an automatic way by the AMG software, from a description of the structure (here a Netlist file generated by PSpice), the commutation mode (cyclic sequence of switched configurations) and the switch control sequence of the static converter to model. This descriptive information is given by the AMG user.

The automatic generation is useful, especially for complex static converter structures. The generated models are also automatically translated into C-S Function, Modelica and VHDL-

AMS language. Indeed, models in such a language are easy to use in the power electronics simulation software like Matlab/Simulink, Portunus, Simplorer, Amesim, Dymola. This also allows to connect the generated model with others by managing automatically the model causality. This aspect is especially very interesting for coupling models.

Intrinsically, for the average modelling, AMG supposes an a priori on the behaviour of the studied static converters. In this way, the user must understand and analyze the behaviour of the static converter to study. It should also be interesting to extract automatically the control and the mode, by analyzing classical simulation.

A specific methodology for the treatment of the discontinuous conduction mode is also proposed with AMG but not presented in this chapter. However, this automatic treatment requires some knowledge about the converter operating in discontinuous conduction mode. Nowadays, it is only automatically applied for DC/DC converters so it will be interesting in the future to extend it to AC/DC and DC/AC converters.

7. Acknowledgment

This work has been supported by the French association ANR, in the context of the project C6E2-SIMPA2 on the simulation and modelling of mechatronic systems.

8. References

Allain, L.; Merdassi, A.; Gerbaud, L. & Bacha, S. (2009). Automatic modelling of Power Electronic Converter, Average model construction and Modelica model generation, *In Proceddings of the Modelica Conference*, Como Italy, September 2009.

Bacha, S.; Brunello, M. & Hassan, A. (1994). A general large signal model for DC-DC symmetric switching converters, *Electric Machines and Power Systems*, Vol 22, N° 4, July 1994 , pp 493-510.

Bass, R.M. & Sun, J. (1998). Using symbolic computation for power electronics, *Computers in Power Electronics*, 6th Workshop, 19-22 July 1998, pp I IV.

Chen, M. & Sun, J. (2005). A General Approach to Averaged Modelling and Analysis of Active-Clamped Converters, Applied Power Electronics Conference and Exposition, *in Proceedings of the Twentieth Annual IEEE* ,Volume 2, 6-10 March 2005, pp 1359 – 1365.

Chetty, P. R. K. (1982). Current injected equivalent circuit approach to modelling and analysis of current programmed switching DC-to-DC converters (discontinuous inductor conduction mode), *IEEE Transactions. on Industry Applications*, Vol. IA-18, N°3, pp 295-299.

Delaye, A.; Albert, L.; Gerbaud, L. & Wurtz, F. (2004). Automatic generation of sizing models for the optimization of electromagnetic devices using reluctances networks, *IEEE Transactions on Magnetics*, Volume 40, Issue2, Part 2, March 2004, pp.830-833.

Etxeberria-Otadui, I.; Manzo, V.; Bacha, S. & Baltes, F. (2002). Generalized average modelling of FACTS for real time simulation in ARENE, *in Proceedings of IEEE-IECON'02*, 5-8 Nov. 2002, Vol.2, pp 864 - 869.

Kanaan, H. Y.; Al-Haddad, K & Fnaiech, F. (2005). Modelling and control of three-phase/switch/level fixed-frequency PWM rectifier: state-space averaged model, *IEEE Proc. Electric Power Applications*, pp. 551-557, May 2005.

Karnopp, D.C.; Margolis, D.L. & Rosenberg, R.C. (1990). System dynamics: a unified approach, *John Wiley and Sons*, 2nd edition.

Kuo-Peng P.; Sdowski, N.; Bastos, J.P.A.; Carlson, R. & Batistela, N.J. (1997). A General Method for Coupling Static Converters with Electromagnetic Structures, *IEEE Transactions on. Magnetics.*, Vol.33, N°2, March1997.

Maksimovic, D.; Stankovic, A.M.; Thottuvelil, V.J. & Verghese, G.C. (2001). Modeling and simulation of power electronic converters, *Proceedings of the IEEE*, Vol. 89, Issue 6, pp 898 – 912.

Maranesi, P.G. & Riva, M. (2003). Automatic modelling of PWM DC-DC converters, *IEEE Power Electron.* Letters, Vol.1, N°4.

Merdassi, A.; Gerbaud, L. & Bacha, S. (2008). A New Automatic Average Modelling Tool for Power Electronics Systems, *IEEE Power Electron. Specialists Conference*, Greece Rhodes, 15-19 June 2008 pp:3425- 3431.

Merdassi, A.; Gerbaud, L. & Bacha, S. (2010). Automatic generation of average models for power electronics systems in VHDL-AMS and Modelica modelling languages, *Journal of Modelling and Simulation of Systems (JMSS)*, Vol.1, Iss.3, pp. 176-186.

Middlebrook, R.D. & Cuk, S. (1976). A general unified approach to modelling switching converter power tages, *IEEE Power Specialists Conference*, pp 18-34, 1976.

Mohan, N.; Underland, T.M. & Robbins, W.P. (1989). Power Electronics: Converters, Applications and Design, 1ere Ed : ISBN: 0 471 61342 82eme :ISBN 0 471 505374, 1989.

Petitclair, P.; Bacha, S. & Rognon, J.p. (1996). Averaged modelling and nonlinear control of an ASVC (advanced static VAr compensator), *Power Electronics Specialists Conference*, 27th Annual IEEE, pp: 753 - 758 vol.1.

Rajagopalan, V. (1987). Computer-aided analysis of power electronic systems, New York, Marcel Dekker Inc.

Sun, J. & Grostollen, H. (1992). Averaged Modeling of switching power converters: Reformulation and theoretical basis, *Power Electronics Specialists Conference PESC Record.*, 23rd Annual IEEE, pp.1165-1172.

Sun, J. & Grostollen, H. (1997). Symbolic Analysis Methods for Averaged Modelling of Switching Power Converters, *IEEE Transactions on Power Electronics*, Vol 12, n° 3, pp 537-546.

Verdiere, F.; Bacha, S. & Gerbaud, L. (2003). Automatic modelling of static converter averaged models, *EPE*, Toulouse, pp 1-9.

Webster, R. & Ngo, K.D.T. (1992). Computer-based symbolic circuit analysis and simulation, *in Conference Proceedings APEC*, pp. 772-779.

7

Robust Control of Active Vehicle Suspension Systems Using Sliding Modes and Differential Flatness with MATLAB

Esteban Chávez Conde[1], Francisco Beltrán Carbajal[2],
Antonio Valderrábano González[3] and Ramón Chávez Bracamontes[3]
[1]*Universidad del Papaloapan, Campus Loma Bonita*
[2]*Universidad Autónoma Metropolitana, Unidad Azcapotzalco, Departamento de Energía*
[3]*Universidad Politécnica de la Zona Metropolitana de Guadalajara*
[4]*Instituto Tecnológico de Cd. Guzmán*
México

1. Introduction

The main control objectives of active vehicle suspension systems are to improve the ride comfort and handling performance of the vehicle by adding degrees of freedom to the system and/or controlling actuator forces depending on feedback and feedforward information of the system obtained from sensors.

Passenger comfort is provided by isolating the passengers from undesirable vibrations induced from irregular road disturbances, and its performance is evaluated by the level of acceleration which vehicle passengers are exposed. Handling performance is achieved by maintaining a good contact between the tire and the road to provide guidance along the track.

The topic of active vehicle suspension control system, for linear and nonlinear models, in general, has been quite challenging over the years and we refer the reader to some of the fundamental work in the area which has been helpful in the preparation of this chapter. Control strategies like Linear Quadratic Regulator (LQR) in combination with nonlinear backstepping control techniques are proposed in (Liu et al., 2006). This strategy requires information about the state vector (vertical positions and speeds of the tire and car body). A reduced order controller is proposed in (Yousefi et al., 2006) to decrease the implementation costs without sacrificing the security and the comfort by using accelerometers for measurements of the vertical movement of the tire and car body. In (Tahboub, 2005) a controller of variable gain that considers the nonlinear dynamics of the suspension system is proposed. It requires measurements of the vertical positions of the car body and the tire, and the estimation of other states and of the road profile.

On the other hand, many dynamical systems exhibit a structural property called differential flatness. This property is equivalent to the existence of a set of independent outputs, called flat outputs and equal in number to the control inputs, which completely parameterizes every state variable and control input (Fliess et al., 1995). By means of differential flatness the analysis and design of controller is greatly simplified. In particular, the combination of differential flatness with sliding modes, which is extensively used when a robust control scheme is required, e.g., parameter uncertainty, exogenous disturbances and un-modeled dynamics (see Utkin, 1978), qualifies as an adequate robust control design approach to get high vibration attenuation level in active vehicle suspension systems. Sliding mode control of a differentially flat system of two degrees of freedom, with vibration attenuation, is presented in (Enríquez-Zárate et al., 2000).

This chapter presents a robust active vibration control scheme based on sliding modes and differential flatness for electromagnetic and hydraulic active vehicle suspension systems. Measurements of the vertical displacements of the car body and the tire are required for implementation of the proposed control scheme. On-line algebraic estimation of the states variables is used to avoid the use of sensors of acceleration and velocity. The road profile is considered as an unknown input disturbance that cannot be measured. Simulation results obtained from Matlab are included to show the dynamic performance and robustness of the active suspension systems with the proposed control scheme. This chapter applies the algebraic state estimation scheme proposed by Fliess and Sira-Ramírez (Fliess & Sira-Ramírez, 2004a, 2004b; Sira-Ramírez & Silva-Navarro, 2003) for control of nonlinear systems, which is based on the algebraic identification methodology of system parameters reported in (Fliess & Sira-Ramírez, 2003). The method is purely algebraic and involves the use of differential algebra. This method is applied to obtain an estimate of the time derivative from any signal, avoiding model reliance of the system at least in the estimation of states. Simulation and experimental results of the on-line algebraic estimation of states on a differentially flat system of two degrees of freedom are presented in (García-Rodríguez, 2005).

This chapter is organized as follows: Section 2 presents the linear mathematical models of vehicle suspension systems of a quarter car. The design of the controllers for the active suspension systems are introduced in Sections 3 and 4. Section 5 divulges the design of the algebraic estimator of states, while Section 6 shows the use of sensors for measuring the variables required by the controller. The simulation results are illustrated in Section 7. Finally, conclusions are brought out in Section 8.

2. Dynamic model of quarter-car suspension systems

2.1 Linear mathematical model of passive suspension system
A schematic diagram of a quarter-car suspension system is shown in Fig. 1(a). The mathematical model of the passive suspension system is given by

$$m_s \ddot{z}_s + c_s(\dot{z}_s - \dot{z}_u) + k_s(z_s - z_u) = 0 \tag{1}$$

$$m_u \ddot{z}_u - c_s(\dot{z}_s - \dot{z}_u) - k_s(z_s - z_u) + k_t(z_u - z_r) = 0 \tag{2}$$

where m_s represents the mass of a quarter car, m_u represents the mass of one wheel with the suspension and brake equipment, c_s is the damper coefficient of suspension, k_s and k_t are the spring coefficients of the suspension and the tire, z_s and z_u are the displacements of car body and the wheel and z_r is the terrain input disturbance. This simplified linear mathematical model of a passive suspension system has been widely used in many previous works, such as (Liu et al., 2006; Yousefi et al., 2006).

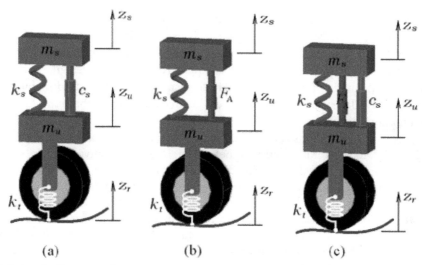

Fig. 1. Schematic diagram of a quarter-vehicle suspension system: (a) passive suspension system, (b) electromagnetic active suspension system and (c) hydraulic active suspension system.

2.2 Linear mathematical model of the electromagnetic active suspension system

A schematic diagram of a quarter-car active suspension system is illustrated in Fig.1 (b). The electromagnetic actuator replaces the damper, forming a suspension with the spring. The friction force of an electromagnetic actuator is neglected. The mathematical model of the electromagnetic suspension system, presented in (Martins et al., 2006), is given by:

$$m_s \ddot{z}_s + k_s(z_s - z_u) = F_A \tag{3}$$

$$m_u \ddot{z}_u - k_s(z_s - z_u) + k_t(z_u - z_r) = -F_A \tag{4}$$

where m_s, m_u, k_s, k_t, z_s, z_u and z_r represent the same parameters and variables shown in the passive suspension system. The electromagnetic actuator force is represented by F_A.

2.3 Linear mathematical model of hydraulic active suspension system

A schematic diagram of an active quarter-car suspension system is shown in Fig. 1(c). The mathematical model of the hydraulic suspension system is given by

$$m_s \ddot{z}_s + c_s(\dot{z}_s - \dot{z}_u) + k_s(z_s - z_u) = -F_f + F_A \tag{5}$$

$$m_u \ddot{z}_u - c_s(\dot{z}_s - \dot{z}_u) - k_s(z_s - z_u) + k_t(z_u - z_r) = F_f - F_A \tag{6}$$

where m_s, m_u, . k_s ., k_t, z_s, z_u and z_r represent the same parameters and variables shown in the passive suspension system. The hydraulic actuator force is represented by F_A, and F_f represents the friction force generated by the seals of the piston with the cylinder wall inside the actuator. This friction force has a significant magnitude ($> 200N$) and cannot be ignored (Martins et al., 2006; Yousefi et al., 2006). The net force given by the actuator is the difference between the hydraulic force F_A and the friction force F_f.

3. Control of electromagnetic suspension system

The mathematical model of the electromagnetic active suspension system illustrated in Fig. 1(b) is given by the equations (3) and (4). Defining the state variables $x_1 = z_s$, $x_2 = \dot{z}_s$, $x_3 = z_u$ and $x_4 = \dot{z}_u$ for the model of the equations mentioned, the representation in the state space form is,

$$\dot{x}(t) = Ax(t) + Bu(t) + Ez_r(t); \quad x(t) \in \mathbb{R}^4, A \in \mathbb{R}^{4 \times 4}, B \in \mathbb{R}^{4 \times 1}, E \in \mathbb{R}^{4 \times 1},$$

$$
\begin{bmatrix} \dot{x}_1 \\ \dot{x}_2 \\ \dot{x}_3 \\ \dot{x}_4 \end{bmatrix} =
\begin{bmatrix}
0 & 1 & 0 & 0 \\
-\dfrac{k_s}{m_s} & 0 & \dfrac{k_s}{m_s} & 0 \\
0 & 0 & 0 & 1 \\
\dfrac{k_s}{m_u} & 0 & -\dfrac{k_s + k_t}{m_u} & 0
\end{bmatrix}
\begin{bmatrix} x_1 \\ x_2 \\ x_3 \\ x_4 \end{bmatrix} +
\begin{bmatrix} 0 \\ \dfrac{1}{m_s} \\ 0 \\ -\dfrac{1}{m_u} \end{bmatrix} u +
\begin{bmatrix} 0 \\ 0 \\ 0 \\ \dfrac{k_t}{m_u} \end{bmatrix} z_r
\tag{7}
$$

with $u = F_A$, the force provided by the electromagnetic actuator as control input.

3.1 Differential flatness

The system is controllable and hence, flat (Fliess et al., 1995; Sira-Ramírez & Agrawal, 2004), with the flat output being the positions of body car and wheel, $F = m_s x_1 + m_u x_3$, in (Chávez-Conde et al., 2009). For simplicity in the analysis of the differential flatness for the suspension system assume that $k_t z_r = 0$. In order to show the parameterization of all the state variables and control input, we firstly compute the time derivatives up to fourth order for F, resulting in

$$F = m_s x_1 + m_u x_3$$
$$\dot{F} = m_s x_2 + m_u x_4$$
$$\ddot{F} = -k_t x_3$$
$$F^{(3)} = -k_t x_4$$
$$F^{(4)} = \frac{k_t}{m_u} u - \frac{k_s k_t}{m_u}(x_1 - x_3) + \frac{k_t^2}{m_u} x_3$$

Then, the state variables and control input are differentially parameterized in terms of the flat output as follows

$$x_1 = \frac{1}{m_s}\left(F + \frac{m_u}{k_t}\ddot{F} \right)$$

$$x_2 = \frac{1}{m_s}\left(\dot{F} + \frac{m_u}{k_t}F^{(3)} \right)$$

$$x_3 = -\frac{1}{k_t}\ddot{F}$$

$$x_4 = -\frac{1}{k_t}F^{(3)}$$

$$u = \frac{m_u}{k_t}F^{(4)} + \left(\frac{k_s m_u}{k_t m_s} + \frac{k_s}{k_t} + 1\right)\ddot{F} + \frac{k_s}{m_s}F$$

3.2 Sliding mode and differential flatness control

The input u in terms of the flat output and its time derivatives is given by

$$u = \frac{m_u}{k_t}F^{(4)} + \left(\frac{k_s m_u}{k_t m_s} + \frac{k_s}{k_t} + 1\right)\ddot{F} + \frac{k_s}{m_s}F \qquad (8)$$

where $F^{(4)} = v$ defines an auxiliary control input. This expression can be written in the following form:

$$u = d_1 F^{(4)} + d_2 \ddot{F} + d_3 F \qquad (9)$$

where $d_1 = \dfrac{m_u}{k_t}$, $d_2 = \dfrac{k_s m_u}{k_t m_s} + \dfrac{k_s}{k_t} + 1$ and $d_3 = \dfrac{k_s}{m_s}$.

Now, consider a linear switching surface defined by

$$\sigma = F^{(3)} + \beta_2 \ddot{F} + \beta_1 \dot{F} + \beta_0 F \qquad (10)$$

Then, the error dynamics restricted to $\sigma = 0$ is governed by the linear differential equation

$$F^{(3)} + \beta_2 \ddot{F} + \beta_1 \dot{F} + \beta_0 F = 0 \qquad (11)$$

The design gains $\beta_2, \ldots, \beta_0$ are selected to verify that the associated characteristic polynomial $s^3 + \beta_2 s^2 + \beta_1 s + \beta_0$ be Hurwitz. As a consequence, the error dynamics on the switching surface $\sigma = 0$ is globally asymptotically stable. The sliding surface $\sigma = 0$ is made globally attractive with the continuous approximation to the discontinuous sliding mode controller as given in (Sira-Ramírez, 1993), i.e., by forcing to satisfy the dynamics,

$$\dot{\sigma} = -\mu[\sigma + \gamma sign(\sigma)] \qquad (12)$$

where μ, γ denote positive real constants and "sign" is the standard signum function. The sliding surface is globally attractive, $\sigma\dot{\sigma} < 0$ for $\sigma \neq 0$, which is a very well known condition for the existence of sliding mode presented in (Utkin, 1978). One then obtains the following sliding-mode controller:

$$u = d_1 v + d_2 \ddot{F} + d_3 F \qquad (13)$$

$$v = -\beta_2 F^{(3)} - \beta_1 \ddot{F} - \beta_0 \dot{F} - \mu[\sigma + \gamma sign(\sigma)]$$

This controller requires the measurement of all the state variables of the suspension system, z_s, $\dot{z}_s$, z_u and $\dot{z}_u$, corresponding to the vertical positions and velocity of the body of the car and the wheel. The variables $\dot{z}_s$ and $\dot{z}_u$ are calculated through an online algebraic estimator, shown in Section 5.

4. Control control of hydraulic suspension system

The mathematical model of the hydraulic active suspension system shown in Fig. 1(c) is given by the equations (5) and (6). Defining the state variables $x_1 = z_s$, $x_2 = \dot{z}_s$, $x_3 = z_u$ and $x_4 = \dot{z}_u$ for the model of the equations mentioned, the representation in the state space form is, $\dot{x}(t) = Ax(t) + Bu(t) + Ez_r(t)$; $x(t) \in \mathbb{R}^4, A \in \mathbb{R}^{4\times4}, B \in \mathbb{R}^{4\times1}, E \in \mathbb{R}^{4\times1}$,

$$
\begin{bmatrix} \dot{x}_1 \\ \dot{x}_2 \\ \dot{x}_3 \\ \dot{x}_4 \end{bmatrix} = \begin{bmatrix} 0 & 1 & 0 & 0 \\ -\dfrac{k_s}{m_s} & -\dfrac{c_s}{m_s} & \dfrac{k_s}{m_s} & \dfrac{c_s}{m_s} \\ 0 & 0 & 0 & 1 \\ \dfrac{k_s}{m_u} & \dfrac{c_s}{m_u} & -\dfrac{k_s+k_t}{m_u} & -\dfrac{c_s}{m_u} \end{bmatrix} \begin{bmatrix} x_1 \\ x_2 \\ x_3 \\ x_4 \end{bmatrix} + \begin{bmatrix} 0 \\ \dfrac{1}{m_s} \\ 0 \\ -\dfrac{1}{m_u} \end{bmatrix} u + \begin{bmatrix} 0 \\ 0 \\ 0 \\ \dfrac{k_t}{m_u} \end{bmatrix} z_r \qquad (14)
$$

with $u = F_A - F_f$, the net force provided by the hydraulic actuator as control input (the net force provided by the actuator is the difference between the hydraulic force F_A and the frictional force F_f).

4.1 Differential flatness

The system is controllable and hence, flat (Fliess et al., 1995; Sira-Ramírez & Agrawal, 2004), with the flat output being the positions of body car and wheel, $F = m_s x_1 + m_u x_3$, in (Chávez-Conde et al., 2009). For simplicity in the analysis of the differential flatness for the suspension system assume that $k_t z_r = 0$. In order to show the parameterization of all the state variables and control input, we firstly compute the time derivatives up to fourth order for F, resulting in

$$
\begin{aligned}
F &= m_s x_1 + m_u x_3 \\
\dot{F} &= m_s x_2 + m_u x_4 \\
\ddot{F} &= -k_t x_3 \\
F^{(3)} &= -k_t x_4 \\
F^{(4)} &= \frac{k_t}{m_u} u - \frac{c_s k_t}{m_u}(x_2 - x_4) - \frac{k_s k_t}{m_u}(x_1 - x_3) + \frac{k_t^2}{m_u} x_3
\end{aligned}
$$

Then, the state variables and control input are parameterized in terms of the flat output as follows

$$
x_1 = \frac{1}{m_s}\left(F + \frac{m_u}{k_t}\ddot{F}\right)
$$

$$
x_2 = \frac{1}{m_s}\left(\dot{F} + \frac{m_u}{k_t}F^{(3)}\right)
$$

$$
x_3 = -\frac{1}{k_t}\ddot{F}
$$

$$
x_4 = -\frac{1}{k_t}F^{(3)}
$$

$$u = \frac{m_u}{k_t}F^{(4)} + \left(\frac{c_s m_u}{k_t m_s} + \frac{c_s}{k_t}\right)F^{(3)} + \left(\frac{k_s m_u}{k_t m_s} + \frac{k_s}{k_t} + 1\right)\ddot{F} + \frac{c_s}{m_s}\dot{F} + \frac{k_s}{m_s}F$$

4.2 Sliding mode and differential flatness control

The input u in terms of the flat output and its time derivatives is given by

$$u = \frac{m_u}{k_t}v + \left(\frac{c_s m_u}{k_t m_s} + \frac{c_s}{k_t}\right)F^{(3)} + \left(\frac{k_s m_u}{k_t m_s} + \frac{k_s}{k_t} + 1\right)\ddot{F} + \frac{c_s}{m_s}\dot{F} + \frac{k_s}{m_s}F \qquad (15)$$

where $F^{(4)} = v$ defines the auxiliary control input. The expression can be written in the following form:

$$u = \eta_1 v + \eta_2 F^{(3)} + \eta_3 \ddot{F} + \eta_4 \dot{F} + \eta_5 F \qquad (16)$$

where $\eta_1 = \frac{m_u}{k_t}$, $\eta_2 = \frac{c_s m_u}{k_t m_s} + \frac{c_s}{k_t}$, $\eta_3 = \frac{k_s m_u}{k_t m_s} + \frac{k_s}{k_t} + 1$, $\eta_4 = \frac{c_s}{m_s}$ and $\eta_5 = \frac{k_s}{m_s}$.

Now, consider a linear switching surface defined by

$$\sigma = F^{(3)} + \beta_2 \ddot{F} + \beta_1 \dot{F} + \beta_0 F \qquad (17)$$

Then, the error dynamics restricted to $\sigma = 0$ is governed by the linear differential equation

$$F^{(3)} + \beta_2 \ddot{F} + \beta_1 \dot{F} + \beta_0 F = 0 \qquad (18)$$

The design gains $\beta_2,\ldots,\beta_0$ are selected to verify that the associated characteristic polynomial $s^3 + \beta_2 s^2 + \beta_1 s + \beta_0$ be Hurwitz. As a consequence, the error dynamics on the switching surface $\sigma = 0$ is globally asymptotically stable. The sliding surface $\sigma = 0$ is made globally attractive with the continuous approximation to the discontinuous sliding mode controller as given in (Sira-Ramírez, 1993), i.e., by forcing to satisfy the dynamics

$$\dot{\sigma} = -\mu[\sigma + \gamma sign(\sigma)] \qquad (19)$$

where μ, γ denote positive real constants and "sign" is the standard signum function. The sliding surface is globally attractive, $\sigma\dot{\sigma} < 0$ for $\sigma \neq 0$, which is a very well known condition for the existence of sliding mode presented in (Utkin, 1978). One then obtains the following sliding-mode controller:

$$u = \eta_1 v + \eta_2 F^{(3)} + \eta_3 \ddot{F} + \eta_4 \dot{F} + \eta_5 F \qquad (20)$$

$$v = -\beta_2 F^{(3)} - \beta_1 \ddot{F} - \beta_0 \dot{F} - \mu[\sigma + \gamma sign(\sigma)]$$

This controller requires the measurement of all the variables of state of suspension system, z_s, $\dot{z}_s$, z_u and $\dot{z}_u$ corresponding to the vertical positions and velocity of the body of the car and the tire, respectively. The variables $\dot{z}_s$ and $\dot{z}_u$ are calculated through an online algebraic estimator, shown in Section 5.

5. On-line algebraic state estimation of active suspension system

5.1 First time derivative algebraic estimation

The algebraic methodology proposed in (Fliess & Sira-Ramírez, 2003) allows us to estimate the derivatives of a smooth signal considering a signal model of $n-th$ order, thus it is not necessary to design the time derivative estimator from a specific dynamic model of the plant. Through valid algebraic manipulations of this approximated model in the frequency domain, and using the algebraic derivation with respect to the complex variable s, we neglect the initial conditions of the signal. The resulting equation is multiplied by a negative power s^{n-1} and returned to the time domain. This last expression is manipulated algebraically in order to find a formula to estimate the first time derivative of $y(t)$.

Consider a fourth order approximation of a smooth signal $y(t)$,

$$\frac{d^4 y(t)}{dt^4} = 0 \tag{21}$$

This model indicates that $y(t)$ is a signal whose behavior can be approximated by a family of polynomials of third order, thus the fourth time derivative is assumed close to zero. The order of this approximation can be increased to enhance the estimation quality of the algebraic estimator. From (21) it is possible to design a time derivative algebraic estimator. First, we apply Laplace transform to (21),

$$s^4 Y(s) - s^3 Y(0) - s^2 \dot{Y}(0) - s\ddot{Y} - Y^{(3)} = 0 \tag{22}$$

Now, taking successive derivatives until a number of three with respect to the complex variable s, we obtain a expression which is free of initial conditions,

$$\frac{d^4 \left(s^4 Y \right)}{ds^4} = 0 \tag{23}$$

Expanding this expression and multiplying by s^{-3},

$$24 s^{-3} Y + 96 s^{-2} \frac{dY}{ds} + 72 s^{-1} \frac{d^2 Y}{ds^2} + 16 \frac{d^3 Y}{ds^3} + s \frac{d^4 Y}{ds^4} \tag{24}$$

Returning to the time domain,

$$\frac{d}{dt}(t^4 z(t)) - 16 t^3 z(t) + 72 \int_0^t \lambda_1^2 z(\lambda_1) d\lambda_1$$
$$- 96 \int_0^t \int_0^{\lambda_1} \lambda_2 z(\lambda_2) d\lambda_2 d\lambda_1 + 24 \int_0^t \int_0^{\lambda_1} \int_0^{\lambda_2} z(\lambda_3) d\lambda_3 d\lambda_2 d\lambda_1 = 0$$

From the last equation is possible to obtain the following algebraic estimator,

$$\frac{dz}{dt} = \frac{12 t^3 z + 96 \int_0^t \int_0^{\lambda_1} \lambda_2 z(\lambda_2) d\lambda_2 d\lambda_1 - 24 \int_0^t \int_0^{\lambda_1} \int_0^{\lambda_2} z(\lambda_3) d\lambda_3 d\lambda_2 d\lambda_1}{t^4} \tag{25}$$

This formula is valid for $t > 0$. Since (25) provides an approximated value of the first derivative, this is only valid during a period of time. So the state estimation should be calculated periodically as follows,

$$\left.\frac{dz}{dt}\right|_{t_i}^{t} = \frac{12(t-t_i)^3 z + 96\int_{t_i}^{t}\int_{t_i}^{\lambda_1}\lambda_2 z(\lambda_2)d\lambda_2 d\lambda_1 \; -24\int_{t_i}^{t}\int_{t_i}^{\lambda_1}\int_{t_i}^{\lambda_2}z(\lambda_3)d\lambda_3 d\lambda_2 d\lambda_1}{(t-t_i)^4} \; , \; \forall (t-t_i) > 0 \tag{26}$$

where (t_i, t) is the estimation period.

In order to obtain a better and smoother estimated value of the vertical velocity, we have implemented simultaneously two algebraic estimators for each velocity to estimate. Proceeding with an out-of-phase policy for one of these algebraic estimators, the outputs of both are combined properly to obtain a final estimated value.

6. Instrumentation of the active suspension system

The only variables required for the implementation of the proposed controllers are the vertical displacements of the body of the car z_s and the vertical displacement of the wheel z_u. These variables are needed to measure by some sensor. In (Chamseddine et al., 2006) the use of sensors in experimental vehicle platforms, as well as in commercial vehicles is presented. The most common sensors used for measuring the vertical displacement of the body of the car and the wheel are laser sensors. This type of sensors could be used to measure the variables z_s and z_s needed for controller implementation. An accelerometer or another type of sensor is not needed to measure the variables $\dot{z}_s$ and $\dot{z}_u$, these variables are estimated with the use of algebraic estimators from knowledge of the variables z_s and z_u. Fig. 2 shows a schematic diagram of the instrumentation for the active suspension system.

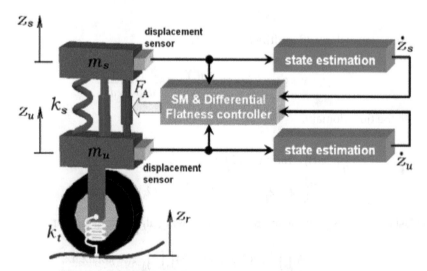

Fig. 2. Schematic diagram of the instrumentation of the active suspension system.

7. Simulation results

The simulation results were obtained by means of MATLAB/Simulink®, with the Runge-Kutta numerical method and a fixed integration step of $1ms$. The numerical values of the quarter-car model parameters (Sam & Hudha, 2006) are presented in Table 1.

Parameter	Value
Sprung mass (m_s)	282 $[kg]$
Unsprung mass (m_u)	45 $[kg]$
Spring stiffness (k_s)	17900 $\left[\dfrac{N}{m}\right]$
Damping constant (c_s)	1000 $\left[\dfrac{N \cdot s}{m}\right]$
Tire stiffness (k_t)	165790 $\left[\dfrac{N}{m}\right]$

Table 1. Quarter-car model parameters

In this simulation study, the road disturbance is shown in Fig. 3 and set in the form of (Sam & Hudha, 2006):

$$z_r = a\frac{1-cos(8\pi t)}{2}$$

with $a = 0.11$ [m] for $0.5 \le t \le 0.75$, $a = 0.55$ [m] for $3.0 \le t \le 3.25$ and 0 otherwise.

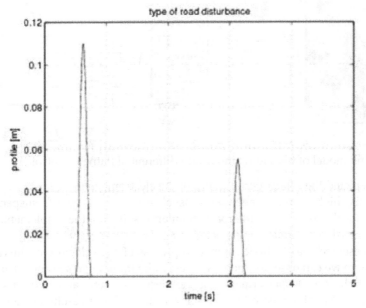

Fig. 3. Type of road disturbance.

It is desired to stabilize the suspension system at the positions $z_s = 0$ and $z_u = 0$. The gains of both electromagnetic and hydraulic suspension controllers were obtained by forcing their closed loop characteristic polynomials to be given by the following Hurwitz polynomial: $(s+p)(s^2 + 2\zeta\omega_n s + \omega_n^2)$ with $p = 100$, $\zeta = 0.5$, $\omega_n = 90$, $\mu = 95$ y $\gamma = 95$.

The Simulink model of the sliding mode and differential flatness controller of the active suspension system is shown in Fig. 4. For the electromagnetic active suspension system, it is assumed that $c_z = 0$. In Fig. 5 is shown the Simulink model of the sliding mode and differential flatness controller with algebraic state estimation.

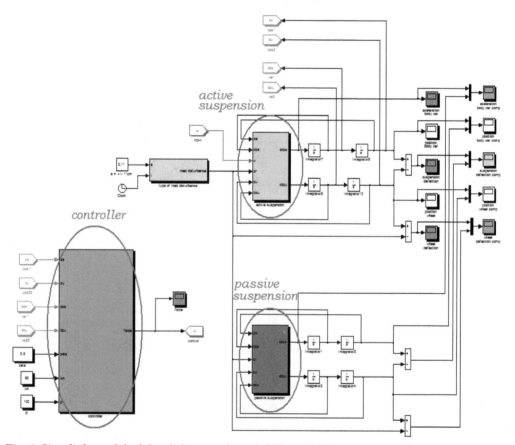

Fig. 4. Simulink model of the sliding mode and differential flatness controller.

In Fig. 6 is depicted the robust performance of the electromagnetic suspension controller. It can be seen the high vibration attenuation level of the active vehicle suspension system compared with the passive counterpart. Similar results on the implementation of the hydraulic suspension controller are depicted in Fig. 7.

In Fig. 8 is presented the algebraic estimation process of the velocities of the car body and the wheel. There we can observe a good and fast estimation. In Figs. 9 and 10 are shown the simulation results on the performance of the electromagnetic and hydraulic suspension controllers using the algebraic estimators of velocities. These results are quite similar to those gotten by the controllers using the real velocities.

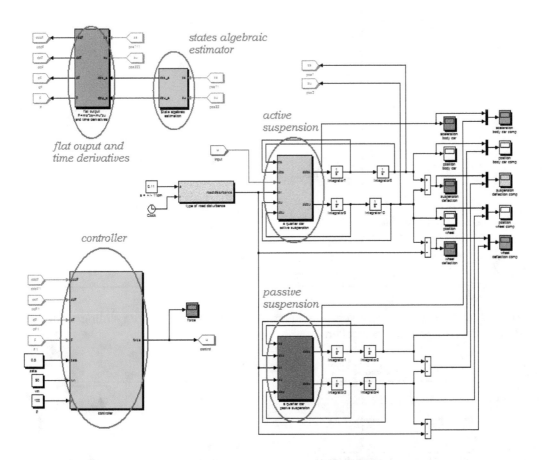

Fig. 5. Simulink model of the sliding mode and differential flatness controller with state estimation.

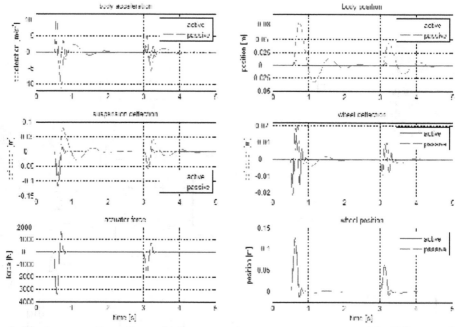

Fig. 6. Electromagnetic active vehicle suspension system responses with sliding mode and differential flatness based controller.

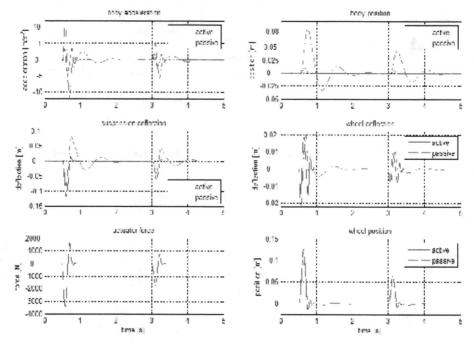

Fig. 7. Hydraulic active vehicle suspension system responses with sliding mode and differential flatness based controller.

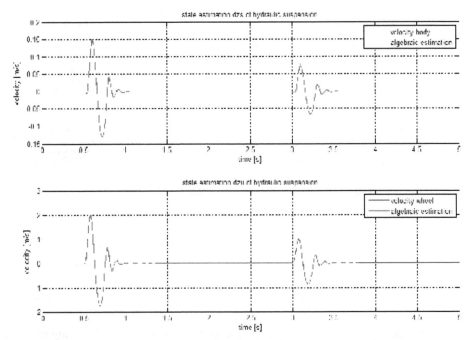

Fig. 8. On-line algebraic state estimates of the hydraulic active suspension system.

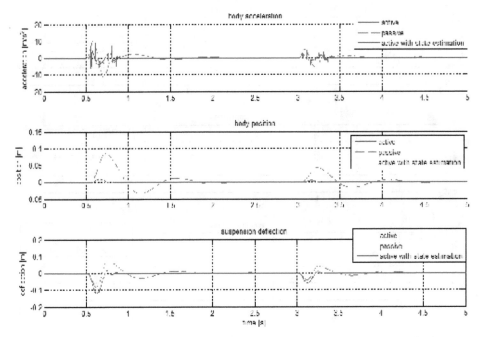

Fig. 9. a. Electromagnetic active vehicle suspension system responses with sliding mode and differential flatness based controller using algebraic state estimation.

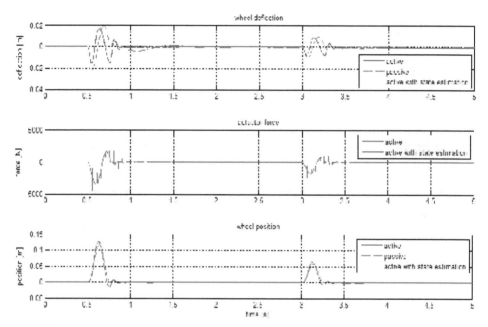

Fig. 9. b. Electromagnetic active vehicle suspension system responses with sliding mode and differential flatness based controller using algebraic state estimation.

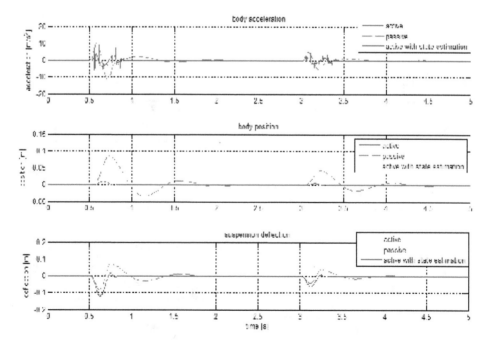

Fig. 10. a. Hydraulic active vehicle suspension system responses with sliding mode and differential flatness based controller using algebraic state estimation.

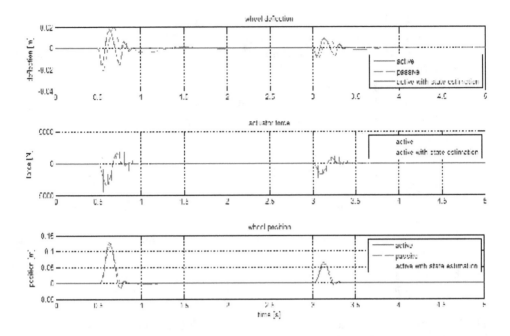

Fig. 10. b. Hydraulic active vehicle suspension system responses with sliding mode and differential flatness based controller using algebraic state estimation.

8. Conclusions

The stabilization of the vertical position of the quarter of car is obtained in a time much smaller to that of the passive suspension system. The sliding mode based differential flatness controller requires the knowledge of all the state variables. Nevertheless the fast stabilization with amplitude in acceleration and speed of the body of the car very remarkable is observed. On-line state estimation is obtained successfully, however when it is used into the controller one can observe a deterioration of the control signal. This can significantly improve with a suitable interpolation between the estimated values at each restart of the integrations. In addition, the simulations results show that the stabilization of the system is obtained before the response of the passive suspension system, with amplitude of acceleration and speed of the body of the car very remarkable. Finally, the robustness of the controllers is observed to take to stabilize to the system before the unknown disturbance.

9. References

Chamseddine, Abbas; Noura, Hassan; Raharijaona, Thibaut "Control of Linear Full Vehicle Active Suspension System Using Sliding Mode Techniques", 2006 IEEE International Conference on Control Applications. pp. 1306-1311, Munich, Germany, October 4-6, 2006.

Chávez-Conde, E.; Beltrán-Carbajal, F.; Blanco-Ortega, A.; Méndez-Azúa, H. "Sliding Mode and Generalized PI Control of Vehicle Active Suspensions", 2009 IEEE International Conference on Control Applications. Saint Petersburg, Russia, July 8-10, 2009.

Enríquez-Zárate, J.; Silva-Navarro, G.; Sira-Ramírez, H. "Sliding Mode Control of a Differentially Flat Vibrational Mechanical System: Experimental Results", 39th IEEE Conference on Decision and Control. pp. 1679-1684, Sydney, Australia, December 2000.

Flies, M.; Lévine, J.; Martin, Ph.; Rouchon, P. "Flatness and defect of nonlinear systems: introductory theory and examples", Int. Journal of Control. Vol. 61, pp. 1327-1361, 1995.

Fliess, M.; Sira-Ramírez, H. "Reconstructeurs d'états". C.R. Acad. Sci. Paris, I-338, pp. 91-96, 2004.

Fliess, M.; Sira-Ramírez, H. "Control via state estimations of some nonlinear systems", 4th IFAC Nolcos Conference. Germany, 2004.

Fliess, M.; Sira-Ramírez, H. "An Algebraic Framework For Linear Identification", ESAIM Control, Optimization and Calculus of Variations. Vol 9, pp. 151-168, January. 2003.

García-Rodríguez, C. "Estimación de estados por métodos algebraicos", Master of Thesis. CINVESTAV-IPN, México, D.F., México, 2005.

Liu, Zhen; Luo, Cheng; Hu, Dewen, "Active Suspension Control Design Using a Combination of LQR and Backstepping", 25th IEEE Chinese Control Conference, pp. 123-125, Harbin, Heilongjiang, August 7-11, 2006.

Martins, I.; Esteves, J.; Marques, D. G.; Da Silva, F. P. "Permanent-Magnets Linear Actuators Applicability in Automobile Active Suspensions", IEEE Trans. on Vehicular Technology. Vol. 55, No. 1, pp. 86-94, January 2006.

Sam, Y. M.; Hudha, K. "Modeling and Force Tracking Control of Hydraulic Actuator for an Active Suspension System", IEEE ICIEA, 2006.

Sira-Ramírez, H. "A dynamical variable structure control strategy in asymptotic output tracking problems", IEEE Trans. on Automatic Control. Vol. 38, No. 4, pp. 615-620, April 1993.

Sira-Ramírez, H.; Silva-Navarro, G. "Algebraic Methods in Flatness, Signal Processing and State Estimation", Innovación Editorial Lagares de México, November 2003.

Sira-Ramírez, H.; Agrawal, Sunil K. "Differentially Flat Systems", Marcel Dekker, N.Y., 2004.

Tahboub, Karim A. "Active Nonlinear Vehicle-Suspension Variable-Gain Control", 13th IEEE Mediterranean Conference on Control and Automation, pp. 569-574, Limassol, Cyprus, June 27-29, 2005.

Utkin,V. I. "Sliding Modes and Their Applications in Variable Structure Systems". Moscow: MIR, 1978.

Yousefi, A.; Akbari, A and Lohmann, B., "Low Order Robust Controllers for Active Vehicle Suspensions", IEEE International Conference on Control Applications, pp. 693-698, Munich, Germany, Octuber 4-6, 2006.

8

Implementation of Induction Motor Drive Control Schemes in MATLAB/Simulink/dSPACE Environment for Educational Purpose

Christophe Versèle, Olivier Deblecker and Jacques Lobry
Electrical Engineering Department, University of Mons
Belgium

1. Introduction

Squirrel-cage induction motors (IM) are the workhorse of industries for variable speed applications in a wide power range that covers from fractional watt to megawatts. However, the torque and speed control of these motors is difficult because of their nonlinear and complex structure. In the past five decades, a lot of advanced control schemes for IM drive appeared. First, in the 1960's, the principle of speed control was based on an IM model considered just for steady state. Therefore, the so-called "scalar control methods" cannot achieve best performance during transients, which is their major drawback. Afterwards, in the 1970's, different control schemes were developed based on a dynamic model of the IM. Among these control strategies, the vector control which is included in the so-called field oriented control (FOC) methods can be mentioned. The principle of vector control is to control independently the two Park components of the motor current, responsible for producing the torque and flux respectively. In that way, the IM drive operates like a separately excited dc motor drive (where the torque and the flux are controlled by two independent orthogonal variables: the armature and field currents, respectively). Since the 1980's, many researchers have worked on improvements of the FOC and vector control which have become the industry's standard for IM drives. Moreover, these researches led to new control strategies such as direct self control (DSC) or direct torque control (DTC). The principle of DTC is to control directly the stator flux and torque of the IM by applying the appropriate stator voltage space vector.

With vector control, FOC, DSC and DTC, the major drawback of the scalar control is overcome because these control schemes are based on a model of the IM which is considered valid for transient conditions (Santisteban & Stephan, 2001). Both DTC and DSC possess high torque dynamics compared to vector control and FOC (Böcker & Mathapati, 2007). However, these two first control techniques have the drawbacks of variable switching frequency and higher torque ripple. The use of space vector pulse-width-modulation (SV-PWM) in conjunction with DTC (called, in this chapter, DTC-space vector modulation or DTC-SVM) has been proposed as a solution to overcome these drawbacks (Rodriguez et al., 2004), but sticking to the fundamental concept of DTC.

The objective of this chapter is neither to do an overview of all IM control methods, as is done, e.g., in (Böcker & Mathapati, 2007; Santisteban & Stephan, 2001), nor to try to ameliorate them. Its aim is to present a powerful tool to help students to understand some IM control schemes, namely the scalar control method, the vector control and the DTC-SVM, of a voltage-fed inverter IM drive using a dSPACE platform and Matlab/Simulink environment. First, they determine the IM parameters using a procedure based on the implemented vector control scheme. The accuracy of these measurements is very important because the considered control methods need precise information about the motor parameters (Böcker & Mathapati, 2007). Then, after concluding off-line simulations, they perform some experiments in speed regulation and speed tracking.

The remainder of this chapter is organized as follows. First, the experimental system is presented. Then the IM models used therein are described and the determination of the IM parameters as well as the implementation of the control methods are exposed. Finally, typical results are presented and the educational experience is discussed.

2. Experimental system

The experimental system consists of three essentials parts: (1) the power driver, (2) the control system and (3) the transducers. Fig. 1 shows a synoptic scheme of the experimental platform.

The power driver consists of a voltage-fed inverter and two machines: one squirrel-type IM of rated power 5.5 kW and a dc-machine with a separately excited field winding of rated power 14.6 kW. These machines are mechanically coupled and the dc-machine is used as the load of the IM.

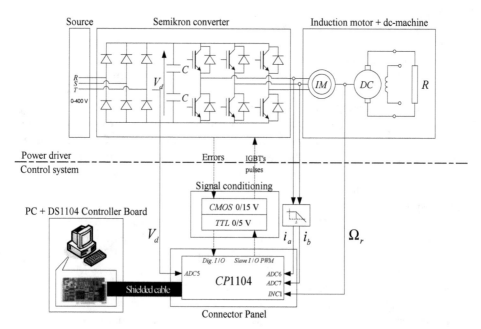

Fig. 1. Synoptic scheme of the experimental platform

The control system is based on the DS1104 Controller Board by dSPACE plugged in a computer. Its development software operates under Matlab/Simulink environment (Bojoi et al., 2002) and is divided into two main components: Real Time Interface (RTI) which is the implementation software and ControlDesk which is the experimentation software. RTI is a Simulink toolbox which provides blocks to configure models (Bojoi et al., 2002). These blocks allow the users to access to the dSPACE hardware. ControlDesk allows, as for it, the users to control and monitor the real-time operation by using a lot of virtual instruments and building a control window.

When using dSPACE, the several steps required to implement a control system on the DS1104 Controller Board are described below. The first step consists in modeling the control system with Simulink and configuring the I/O connections of the Connector Panel thanks to the RTI toolbox. After that, the Real-Time Workshop (RTW) toolbox, using RTI, automatically generates the C-code for the board. Once the execution code has been generated, the dSPACE hardware can perform a real-time experiment which can be controlled from a PC with ControlDesk. ControlDesk can be used to monitor the simulation progress, adjust parameters online, capture data (in a format compatible with Matlab) and communicate easily with the upper computer real-time (Luo et al., 2008). Fig. 2 presents the connections between Matlab and dSPACE.

The experimental system contains several current and voltage LEM transducers as well as a speed sensor. Moreover, the measured currents must be filtered in order to avoid aliasing when they will be converted into digital signals. Therefore, an anti-aliasing filter is added to each current transducer. The cut-off frequency of this filter is estimated at 500 Hz (an order of magnitude above the rated frequency of 50 Hz).

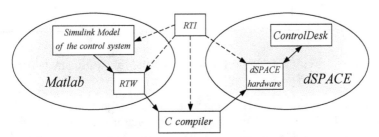

Fig. 2. Connections between Matlab and dSPACE (Mäki et al., 2005)

3. IM modeling

A simple per phase equivalent circuit model of an IM is interesting for the analysis and the performance prediction at steady-state condition. Such a model is therefore used in this chapter for scalar control. In the two other considered control schemes, namely vector control and DTC-SVM, the transient behaviour of the IM has to be taken into account. So, the dynamic d-q model of the IM based on the Park's transformation is considered. These two well-known types of IM models are available in literature (see, e.g., (Bose, 2002; Mohan, 2001)) and are briefly recalled in this section.

3.1 Steady-state model of IM

The considered steady-state IM model, based on complex phasors, is shown in Fig. 3(a). This model is an equivalent circuit with respect to the stator and can easily be established from the short-circuited transformer-equivalent circuit (Bose, 2002).

The synchronously rotating air gap flux wave generates an electromotive force E_s which differs from the stator terminal voltage V_s by the drop voltage in stator resistance R_s and stator leakage inductance l_s. The stator current I_s consists of the magnetizing current I_m and the rotor current I_r (referred to the stator in Fig. 3(a)) which is given by (Bose, 2002):

$$I_r = \frac{E_s}{\frac{R_r}{s} + j \cdot 2 \cdot \pi \cdot f \cdot l_r}. \tag{1}$$

where R_r is the rotor resistance (referred to the stator), l_r is the rotor leakage inductance (referred to the stator), s is the slip and f is the stator supply frequency. Finally, in Fig. 3(a), L_m is the magnetizing inductance and R_m is the equivalent resistance for the core loss.

The equivalent circuit of Fig. 3(a) can be further simplified as shown in Fig. 3(b) where the resistance R_m has been dropped and the inductance l_s has been shifted to the other side of the inductance L_m.

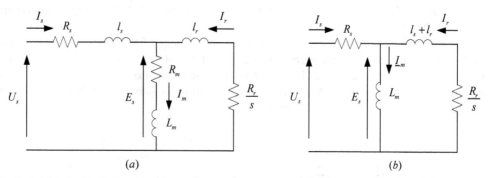

(a)   (b)

Fig. 3. (a) Equivalent circuit with respect to the stator and (b) Approximate per phase equivalent circuit

This approximation is easily justified as the value of the magnetizing inductance is, at least, one order of magnitude greater than the value of the stator leakage inductance. Moreover, the performance prediction by the approximate equivalent circuit typically varies within 5% from that of the actual machine (Bose, 2002).

Based on the equivalent circuit of Fig. 3(b), it can be demonstrated (see, e.g., (Bose, 2002)) the following expression of the electromagnetic torque:

$$T_e = 3 \cdot P \cdot R_r \cdot \frac{(s \cdot \omega) \cdot \psi^2}{R_r^2 + (s \cdot \omega)^2 \cdot (l_s + l_r)^2} \tag{2}$$

where P is the number of poles pairs and ψ is the stator flux. It should also be emphasized that for small slip values, (2) can be rewritten as:

$$T_e = \frac{3 \cdot P \cdot \psi^2}{R_r} \cdot (s \cdot \omega) = K_T \cdot (s \cdot \omega) \tag{3}$$

where K_T is a constant if the flux is kept constant.

Finally, for steady-state operation of IM, the rotational speed Ω_r can be calculated by:

$$\Omega_r = (1 - s) \cdot \frac{2 \cdot \pi \cdot f}{P}. \tag{4}$$

3.2 Dynamic model of IM

Applying the usual space vector transformation to a three-phase system, it is possible to obtain the following set of equations, based on space vectors, that describes the IM dynamic behavior in a stator fixed coordinate system (Rodrigez et al., 2004):

$$\mathbf{v}_s = R_s \cdot \mathbf{i}_s + \frac{d\mathbf{\psi}_s}{dt} + j \cdot \omega_1 \cdot \mathbf{\psi}_s \tag{5}$$

$$\mathbf{v}_r = 0 = R_r \cdot \mathbf{i}_r + \frac{d\mathbf{\psi}_r}{dt} + j \cdot (\omega_1 - P \cdot \Omega_r) \cdot \mathbf{\psi}_r \tag{6}$$

$$\mathbf{\psi}_s = l_s \cdot \mathbf{i}_s + L_m \cdot (\mathbf{i}_s + \mathbf{i}_r) \tag{7}$$

$$\mathbf{\psi}_r = l_r \cdot \mathbf{i}_r + L_m \cdot (\mathbf{i}_s + \mathbf{i}_r) \tag{8}$$

where ω_1 is the rotational speed of the d-q reference frame, $\mathbf{v}_s$ and $\mathbf{v}_r$ are respectively the stator and rotor voltages, $\mathbf{i}_s$ and $\mathbf{i}_r$ are respectively the stator and rotor currents and $\mathbf{\psi}_s$ and $\mathbf{\psi}_r$ are respectively the stator and rotor fluxes.

The electromagnetic torque developed by the IM can be expressed in terms of flux and current space vector by several analog expressions. In this chapter, two of them are useful:

$$T_e = \frac{3}{2} \cdot P \cdot \frac{L_m}{L_r} \cdot [\mathbf{\psi}_r \times \mathbf{i}_s] \tag{9}$$

and

$$T_e = \frac{3}{2} \cdot P \cdot \frac{L_m}{\sigma \cdot L_s \cdot L_r} \cdot [\mathbf{\psi}_s \times \mathbf{\psi}_r] \tag{10}$$

where L_s ($L_s = L_m + l_s$) and L_r ($L_r = L_m + l_r$) are respectively the stator and rotor inductances and σ is the leakage factor of the IM defined as:

$$\sigma = 1 - \frac{L_m^2}{L_s \cdot L_r} . \tag{11}$$

The speed Ω_r cannot normally be treated as a constant (Bose, 2002). It can be related to torque using the IM drive mechanical equation which is:

$$T_e = J \cdot \frac{d\Omega_r}{dt} + B \cdot \Omega_r \tag{12}$$

where J and B denote respectively the total inertia and total damping of the IM coupled to the dc-machine.

Finally, the dynamic model of the IM is shown in Fig. 4 in its complex form for compact representation.

4. Control schemes

4.1 Scalar control

Scalar control is due to magnitude variation of the control variables only and disregards the coupling effect in the machine (Bose, 2002). Although having inferior performances than the

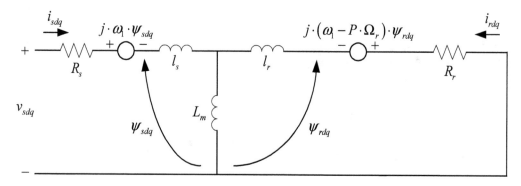

Fig. 4. Dynamic model of IM

two other considered control schemes, it is considered in this chapter because it has been widely used in industry and it is easy to implement. So, this is a good basis for the study of IM drive. Among the several scalar control techniques, the open loop Volts/Hz control of IM, which is the most popular method of speed control of a voltage-fed inverter IM, is used. For adjustable speed applications, frequency control appears to be natural considering (4). However, the stator voltage is required to be proportional to frequency so that the flux:

$$\psi = \frac{U_s}{2 \cdot \pi \cdot f} \tag{13}$$

remains constant (normally equal to its rated value), neglecting the stator resistance voltage drop. Indeed, if an attempt is made to reduce the frequency at the rated stator voltage, the flux will tend to saturate, causing excessive stator current and distortion of the flux waveforms (Bose, 2002). Therefore, in the region below the base frequency f_b (which can be different from the rated frequency), a reduction of the frequency should be accompanied by the proportional reduction of the stator voltage in order to maintain the flux to its rated value. Fig. 5 shows the torque-speed curves. It can be shown in this figure that in the region below the base frequency, the torque-speed curves are identical to each other and that the maximum torque T_{em} (which can be deduced from (2)) is preserved regardless of the frequency, except in the low-frequency region where the stator resistance voltage drop must be compensated by an additional boost voltage U_{boost} in order to restore the T_{em} value (as shown in Fig. 5).

Once the frequency increases beyond the base frequency, the stator voltage is kept constant and, therefore, the flux decreases. As shown in Fig. 5, in this flux-weakening region, the torque-speed curves differ from the previous ones and the developed torque decreases. An advantage of the flux-weakening mode of operation is that it permits to increase the speed range of the IM.

Fig. 6 shows the block diagram of the Volts/Hz speed control method. The frequency f^*, which is approximately equal to the speed neglecting the slip, is the primary control variable. The secondary control variable is the stator voltage U_s^* which is directly generated from the frequency command using the gain factor G so that the flux remains constant or a flux-weakening mode of operation is achieved. Moreover, at low-frequencies, corresponding to a low rotational speed of the IM, the boost voltage is added in order to have the maximum torque available at zero speed. Note that the effect of this boost voltage becomes negligible when the frequency grows. Finally, the f^* command signal is integrated to generate the angle signal θ^* which in combination with U_s^* produce the sinusoidal voltages.

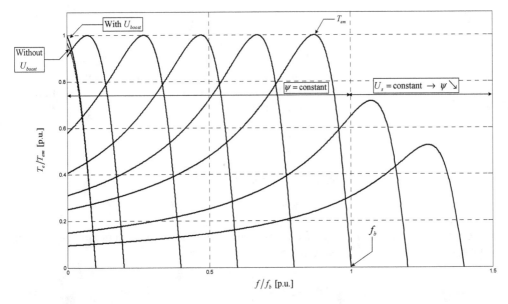

Fig. 5. Torque-speed curves at constant Volts/Hz region and in flux-weakening region

Note that an improvement of the open-loop Volts/Hz control is the close-loop speed control by slip regulation (see, e.g., (Bose, 2002)).

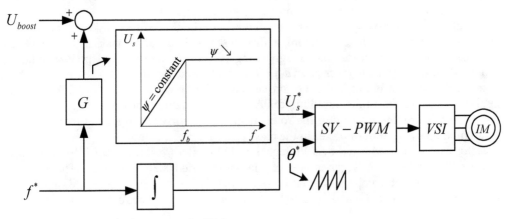

Fig. 6. Open-loop Volts/Hz control of IM

4.2 Vector control
The objective of vector control of IM is to allow an IM to be controlled just like a separately excited dc motor drive (where the torque and the flux are controlled by two independent orthogonal variables: the armature and field currents, respectively). This is achieved by a

proper choice of the so-called Park's rotating frame (d-q axes) in which the space vector equations (5)-(8) are separated into d-q ones.

In the d-q frame, (9) can be rewritten:

$$T_e = \frac{3}{2} \cdot P \cdot \frac{L_m}{L_r} \cdot \left(\psi_{rd} \cdot i_{sq} - \psi_{rq} \cdot i_{sd} \right) \tag{14}$$

where ψ_{rd} and ψ_{rq} are the rotor fluxes and i_{sd} and i_{sq} are the stator current in the dq-axes.

In this chapter, a direct vector control strategy in which the d-q frame rotates along with the rotor flux (which is maintained at its rated value) is considered. The d-axis is aligned with the direction of the rotor flux. Therefore, the q-axis component of the rotor flux is null and the expression of the electromagnetic torque simplifies as follows:

$$T_e = \frac{3}{2} \cdot P \cdot \frac{L_m^2}{L_r} \cdot i_{mr} \cdot i_{sq} \tag{15}$$

where i_{mr} is the rotor magnetizing current defined by:

$$\psi_{rd} = L_m \cdot i_{mr} \cdot \tag{16}$$

Using (6) and (7) projected in the d-q frame as well as (16), the rotor magnetizing current can be expressed in terms of the d-axis stator current as follows:

$$i_{mr} + T_r \cdot \frac{di_{mr}}{dt} = i_{sd} \tag{17}$$

where T_r ($T_r = L_r/R_r$) is the rotor time constant.

From (15) and (17), one can conclude that the d-axis stator current (i_{sd}) is controlled to maintain the flux at its rated value whereas the q-axis stator current (i_{sq}) is varied to achieve the desired electromagnetic torque. Therefore, the IM can be controlled just like a separately excited dc motor drive because the d- and q-axes are orthogonal.

Note that, at each sample time, direct vector control of the IM requires to know the module and phase of the rotor flux. Therefore, a flux observer is used. Its task is to provide an estimate of the rotor flux (or rotor magnetizing current) in module and phase using the measured stator currents (converted in d-q components) and speed.

Once a flux estimate is available, the torque can easily be computed using (15). The flux observer implemented in the Simulink/dSPACE environment is shown in Fig. 7. As can be seen in this figure, the phase μ of the rotor magnetizing current is obtained by integrating the angular speed ω_1 of the d-q axes in the fixed stator reference frame.

In vector control, the IM is fed by a Voltage Supply Inverter (VSI) and a SV-PWM is used to produce the instantaneous generation of the commanded stator voltages. The required d-axis and q-axis stator voltages (v_{sd} and v_{sq}) that the VSI must supply to the IM, in order to make the stator currents (i_{sd} and i_{sq}) equal to their reference values are expressed as follows:

$$v_{sd} = R_s \cdot \left[i_{sd} + T_s \cdot \left(\sigma \cdot \frac{di_{sd}}{dt} + (1-\sigma) \cdot \frac{di_{mr}}{dt} - \sigma \cdot \omega_1 \cdot i_{sq} \right) \right] \tag{18}$$

$$v_{sq} = R_s \cdot \left[i_{sq} + T_s \cdot \left(\sigma \cdot \frac{di_{sq}}{dt} + (1-\sigma) \cdot \omega_1 \cdot i_{mr} - \sigma \cdot \omega_1 \cdot i_{sd} \right) \right] \tag{19}$$

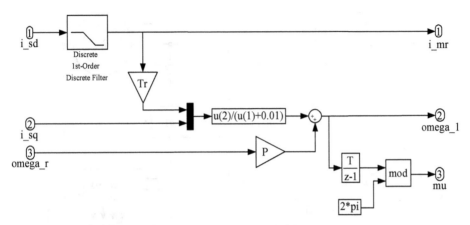

Fig. 7. Flux observer implemented in Matlab/Simulink/dSPACE environment

where T_s ($T_s = L_s/R_s$) is the stator time constant.

In the d-axis (respectively q-axis) voltage equation, only the first two terms of the right-hand side are due to the d-component (respectively q-component) of the stator current, i_{sd} (respectively i_{sq}). The other terms, due to i_{mr} and i_{sq} (respectively i_{mr} and i_{sd}) can be considered as disturbances (Mohan, 2001). So, (18) and (19) can simply be rewritten as:

$$v'_{sd} = R_s \cdot \left[i_{sd} + \sigma \cdot T_s \cdot \frac{di_{sd}}{dt} \right] \tag{20}$$

$$v'_{sq} = R_s \cdot \left[i_{sq} + \sigma \cdot T_s \cdot \frac{di_{sq}}{dt} \right]. \tag{21}$$

Note that, in order to account for the terms considered as disturbances, a "decoupling compensator" block is incorporated in the vector control scheme.

The vector control scheme is shown in Fig. 8 with the reference values indicated by '*'. This control scheme matches to a direct vector control strategy in which the rotor flux is assigned to a reference value (its rated value as mentioned above).

The d-axis reference current, i^*_{sd}, controls the rotor flux (through the rotor magnetizing current) whereas the q-axis current, i^*_{sq}, controls the electromagnetic torque developed by the IM. The reference currents in the d- and q-axes are generated by a flux control loop and a speed control loop, respectively. The d-axis and q-axis voltages, v'_{sd} and v'_{sq}, are calculated from the given reference currents and using (20) and (21). To obtain these command signals, two PI controllers (in two inner current loops) are employed and it is assumed that the compensation is perfect. The terms considered as disturbances are then added to the voltages v'_{sd} and v'_{sq} in the "decoupling compensator" block. Hence, the d-axis and q-axis reference voltages, v_{sd} and v_{sq}, are obtained. Finally, these reference voltages are converted into the reference three-phase voltages v_a, v_b and v_c and supplied (in an approximate fashion over the switching period) by the VSI using the SV-PWM technique. Note that the four controllers are tuned by the pole-zero compensation technique.

4.3 DTC-SVM

Direct torque control of IM consists in closed-loop control of the stator flux and torque using estimators of these two quantities (Trzynadlowski et al., 1999).

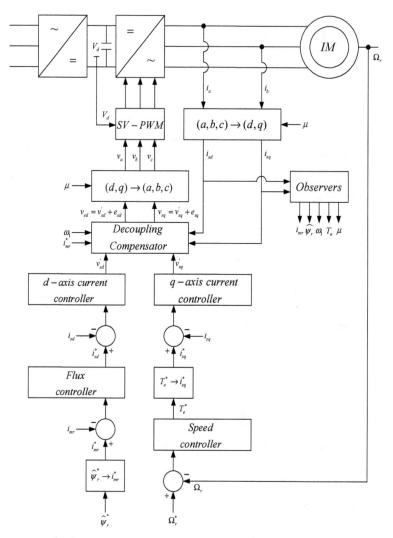

Fig. 8. Vector control scheme

From the IM dynamic model (5)-(8), the following equation can be obtained:

$$\frac{d\psi_r}{dt} + \left(\frac{1}{\sigma \cdot T_r} - j \cdot P \cdot \Omega_r\right) \cdot \psi_r = \frac{L_m}{\sigma \cdot T_r \cdot L_s} \cdot \psi_s \tag{22}$$

which shows that the relationship between the stator and rotor fluxes is of low-pass filter type with time constant $\sigma \cdot T_r$ (Casadei et al., 2005). Otherwise stated, the rotor flux will follow a change in the stator flux with some delay (typically of the order of a few tenths of ms).

Equation (10) can also be rewritten as follows:

$$T_e = \frac{3}{2} \cdot P \cdot \frac{L_m}{\sigma \cdot L_s \cdot L_r} \cdot \psi_s \cdot \psi_r \cdot \sin\delta \tag{23}$$

where δ is the angle between the stator and rotor fluxes.

Based on the previous expression, it is clear that it is possible to achieve machine torque control directly by actuating over the angle δ (Rodriguez et al., 2004). DTC is founded on this consideration.

Due to slow rotor flux dynamic, the easiest way to change δ is to force a variation in the stator flux. Neglecting the effect of the voltage drop across the stator resistance, the stator flux vector in (5) is the time integral of the stator voltage vector. Hence, for sampling time Δt sufficiently small, (5) can be approximated as (Rodriguez et al., 2004):

$$\Delta \psi_s \approx \Delta t \cdot v_s \qquad (24)$$

which means that the stator flux can be changed in accordance with the stator voltage vector supplied to the IM.

In the conventional DTC drive, the flux and torque magnitude errors are applied to hysteresis comparators. The outputs of these comparators are used to select the appropriate stator voltage vector to apply to the IM by means of a pre-designed look-up table. This scheme is simple and robust but it has certain drawbacks such as variable switching frequency and large current and torque ripples. Note that these ripples are imposed by the hysteresis-band width chosen for the hysteresis comparators.

In this chapter, the conventional DTC scheme is not considered but the so-called DTC-SVM scheme, which is discussed in details in (Rodriguez et al., 2004) and briefly presented below, is used. The objective of this alternative DTC scheme is to select the exact stator voltage vector that changes the stator flux vector according to the δ angle reference (calculated from the torque reference), while keeping the stator flux magnitude constant. SV-PWM technique is used to apply the required stator voltage vector to the IM.

The DTC-SVM scheme is shown in Fig. 9 with the reference values once again indicated by '*'. In this scheme, the torque reference is obtained from the speed reference thanks to the same speed controller as the one used in vector control. The error between the estimated torque and its reference value is processed through a PI controller to calculate the δ angle. From this angle, as well as the module of the stator flux (imposed at its rated value) and the estimated phase of the rotor flux, the block "Flux calculator" gives the stator flux reference according to:

$$\psi_s^* = \left| \psi_s \right|^* \cdot \cos(\delta + \angle \psi_r) + j \cdot \left| \psi_s \right|^* \cdot \sin(\delta + \angle \psi_r) \qquad (25)$$

The error between this reference quantity and the estimated stator flux is then divided by Δt, according to (24), in order to obtain the reference stator voltage vector. An approximation of that vector (over the switching period) is then supplied to the IM by the VSI using SV-PWM.

As explained above, the stator and rotor fluxes as well as the torque must be estimated. This is realized by the block "Torque and flux estimators". The rotor flux estimator is based on the IM model and implemented as follows:

$$\psi_r = \frac{1}{T_r} \int \left(L_m \cdot i_s - (1 - j \cdot T_r \cdot P \cdot \Omega_r) \cdot \psi_r \right) \cdot dt . \qquad (26)$$

The stator flux is, as for it, derived from:

$$\psi_s = \sigma \cdot L_s \cdot i_s + \frac{L_r}{L_m} \cdot \psi_r . \qquad (27)$$

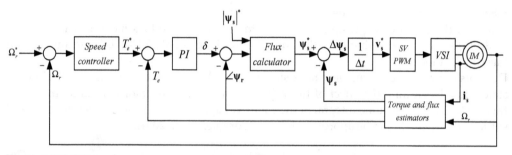

Fig. 9. DTC-SVM scheme

Once the stator and rotor fluxes estimates have been made available, the torque can be easily computed using (23). The observers of the stator and rotor fluxes implemented in the Matlab/Simulink/dSPACE environment are shown in Fig. 10.

The speed controller is tuned by the pole-zero compensation technique whereas the PI, that generates the δ angle reference, is tuned through a method based on the relay feedback test (Padhy & Majhi, 2006).

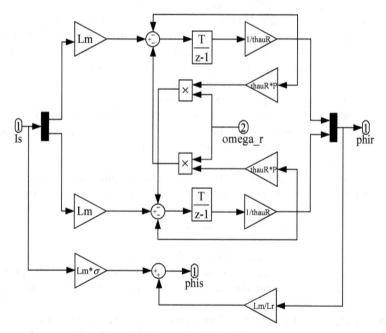

Fig. 10. Fluxes observers implemented in Matlab/Simulink/dSPACE environment

5. Determination of IM parameters

As it is well known, the control methods of IM require an accurate determination of machine parameters. Typically, five parameters need to be determined:

- R_s, R_r – the stator and rotor resistances;
- l_s, l_r – the stator and rotor leakage inductances;
- L_m – the magnetizing inductance.

In this chapter, the above parameters are determined both by a procedure founded on the vector control scheme (discussed in details in (Bose, 2002; Khambadkon et al., 1991)) and by a conventional method (no-load and blocked rotor tests).

The method based on the vector control scheme is divided into eight steps. Initially, the name plate machine parameters are stored into the computer's memory. Precisely, the method needs to know the rated stator voltage and current, the rated frequency and the number of poles pairs of the IM.

In a second step, the stator resistance is determined by a test at dc level. Next, the stator transient time constant is measured. Thanks to the stator resistance and transient time constant, the two inner current loops of the vector control scheme can be tuned. In the fifth step, the rotor time constant is determined. Next, the magnetizing inductance is measured. Note that, in this sixth step, the value of the rotor time constant is adjusted online in order to achieve $i_{sd} = i_{mr}$. Therefore, its value is verified because it is the most important parameter to obtain accurate rotor flux estimation. Thanks to the value of the rotor time constant, the flux controller of the vector control scheme can be tuned in the seventh step. Finally, the mechanical parameters of the IM coupled with the dc-machine (see below) are determined and the speed controller can be tuned.

The method based on the vector control structure has the major advantage that it can be automated. Therefore, this procedure can evaluate the motor parameters at each start-up of the experimental platform. All the results are reported in Table 1. As can be seen, very small differences are obtained between the two approaches, whatever the parameters. Therefore, one can conclude that all the IM parameters are properly determined.

Moreover, the procedure based on the vector control structure permits to determine the mechanical parameters of the IM coupled with the dc-machine. Three parameters need to be determined:

- J – the total inertia;
- T_m – the mechanical time constant;
- B – the total damping factor.

All the results are reported in Table 2.

Parameters	Conventional method	Method based on the vector control scheme
R_s	0.81 Ω	1.26 Ω
R_r	0.57 Ω	0.42 Ω
l_s	4 mH	3.9 mH
l_r	4 mH	3.9 mH
L_m	160 mH	164 mH

Table 1. IM parameters

Parameters	Values
J	0.0731 kg·m²
T_m	2.1 s
B	0.00348 kg·m²/s

Table 2. Mechanical parameters

6. Experimental results

After concluding off-line simulations using the blocks of the SimPowerSystem toolbox in Simulink software, the three control schemes are tested and compared by means of speed reference tracking, torque dynamic and flux regulation. To do so, the validated control strategies are transferred to the digital control board. Fig. 11 and 12, respectively, show the vector control algorithm implemented with Simulink and the corresponding window of the ControlDesk software which controls the dSPACE hardware. Note that more details about this work can be found in (Versèle et al., 2008; Versèle et al. 2010).

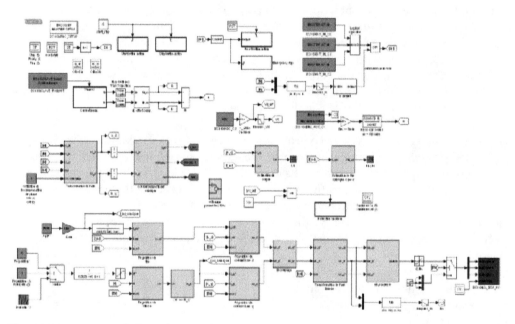

Fig. 11. Vector control algorithm implemented with Simulink

After the algorithms implementation, several tests can be carried out in order to plot the torque-speed curves in scalar control as well as to evaluate the vector control and DTC-SVM behavior in speed regulation and speed tracking. For all these tests, the switching frequency of the VSI is set at 9 kHz. The sample frequency is chosen equal to the switching frequency.

6.1 Scalar control

Fig. 13 shows the predetermined torque-speed curves (solid line) for several supplied frequencies in scalar control (in the constant flux region and in the flux-weakening region) as well as the measured torque-speed curves (circles). Note that the base frequency has been arbitrarily chosen equal to 40 Hz. It should also be noticed that, for the smaller frequencies (viz. 30 Hz, 35 Hz and 40 Hz), the measures have been stopped when instability, which can appear at low-speed and light-load operation (Kishimoto et al., 1986), is observed. One can easily conclude a very good agreement between the predetermined torque-speed curves and the measures. This confirms that the parameters are properly determined.

Fig. 14 shows the torque-speed curves in scalar control and constant rotor frequency $(s \cdot f)$ operation. Note that this rotor frequency is maintained equal to 1 Hz by adjusting the slip by

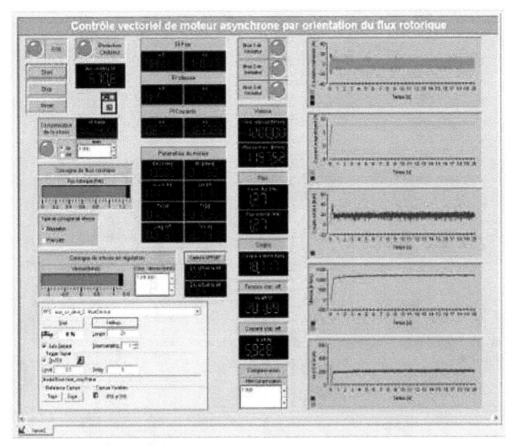

Fig. 12. ControlDesk virtual control panel

acting on the excitation of the dc-machine. Once again, one can conclude a good agreement between the predetermined curve (in solid line) and the measures performed from 30 Hz to 60 Hz (circles). Moreover, considering (3), the results can easily be explained. Indeed, in the constant flux region (below the base speed Ω_{base}), the torque is constant as the rotor frequency and the flux are constant (see (3)). Then, in the flux-weakening region, the rotor frequency remains constant but the flux decreases in inverse proportion to the frequency (and, so, to the speed). Therefore, according to (3), the torque decreases in inverse proportion to the square of the speed.

Finally, some measures have been performed in close-loop speed control with Volts/Hz control and slip regulation. Due to the slip regulation, the IM is able to rotate at the synchronous speed (900 rpm at 30 Hz and 1500 rpm at 50 Hz as $P = 2$) and, therefore, the torque-speed curves become vertical lines as shown in Fig. 15. As a final point, Fig. 16 shows a linear evolution of the torque as a function of the rotor frequency which can be explained on the basis of (3). Moreover, the slope of the regression lines, viz. 29.29 Nm/Hz at 30 Hz and 16.34 Nm/Hz at 50 Hz, are very close to the theoretical values of K_T (see (3)), viz. 28.92 Nm/Hz at 30 Hz and 16.27 Nm/Hz at 50 Hz. One can also conclude that the statistics of the regression, i.e. the R^2 and R^2_{Adj}, are very good (close to 1 and close to each other).

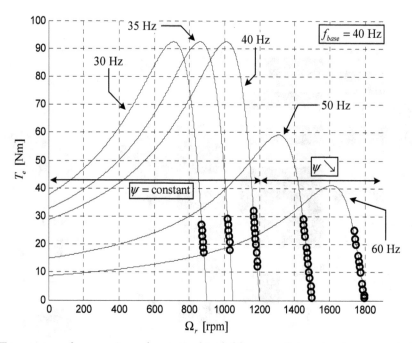

Fig. 13. Torque-speed curves in scalar control (solid line: predetermined curves; circles: measures)

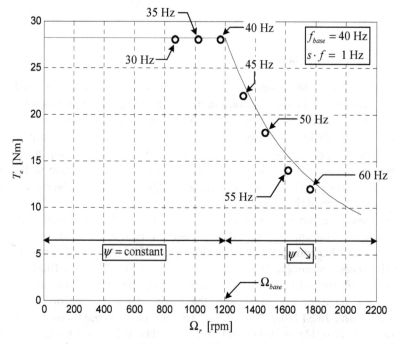

Fig. 14. Torque-speed curves in scalar control and constant rotor frequency (solid line: predetermined curves; circles: measures)

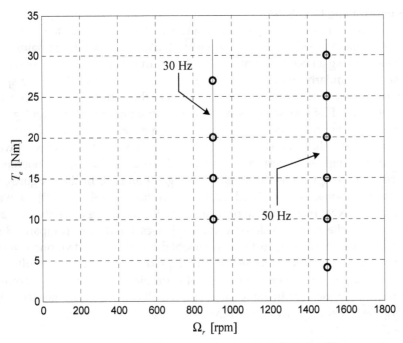

Fig. 15. Torque-speed curves in close-loop speed control with Volts/Hz control and slip regulation

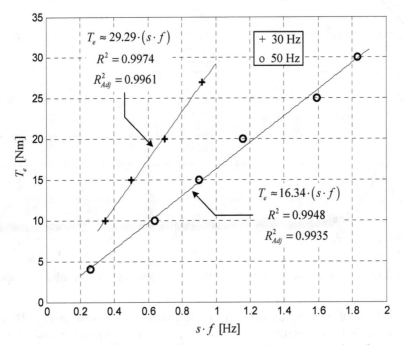

Fig. 16. Evolution of the torque as a function of the rotor frequency in close-loop speed control with Volts/Hz control and slip regulation

6.2 Vector control and DTC-SVM

In this sub-section, the results in vector control and DTC-SVM are presented and discussed. First, a speed regulation test is carried out at no-load (the load generator remains coupled to the shaft) by supplying a speed reference of 1000 rpm. In Fig. 17, the response using vector control is shown. Similarly, Fig. 18 shows the response with DTC-SVM. The speed, flux magnitude and torque are represented (from the top to the bottom). One can conclude that vector control and DTC-SVM schemes have similar dynamic responses and good speed reference tracking. One can also observe that, in vector control scheme, the rotor flux is correctly regulated and that, in DTC-SVM, the stator flux is properly regulated as well.

Then, a speed tracking test is performed at no-load. To do so, the speed reference varies from 1000 rpm to -1000 rpm as can be seen in Fig. 19 and 20. In these figures, one can observe a good response to the speed profile in vector control (Fig. 19) as well as in DTC-SVM (Fig. 20). Note that, as in the previous tests, the fluxes are correctly regulated.

In this sub-section, the two considered control schemes for IM are compared in similar operating conditions. According to the experimental results presented, one can conclude that vector control and DTC-SVM of IM drive produce comparable results in speed regulation and tracking. However, the DTC-SVM is simplest and easiest to implement than vector control. Indeed, no coordinate rotation and less PI controllers are needed. Moreover, less torque ripple is observed with DTC-SVM compare to vector control.

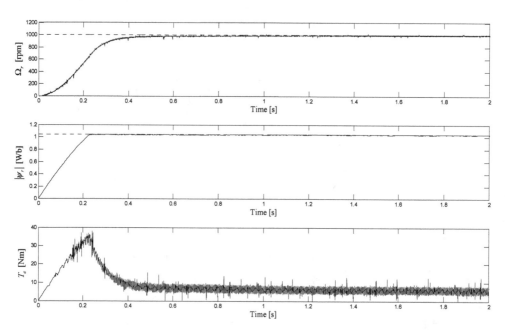

Fig. 17. Speed regulation in vector control (dotted line: speed reference; solid line: speed measured)

Finally, the effects of a load variation in vector control are shown in Fig. 21. One can conclude that such load variation does not affect the flux regulation and that, after a transient phase, the speed remains regulated at the considered speed reference (viz. 1200 rpm).

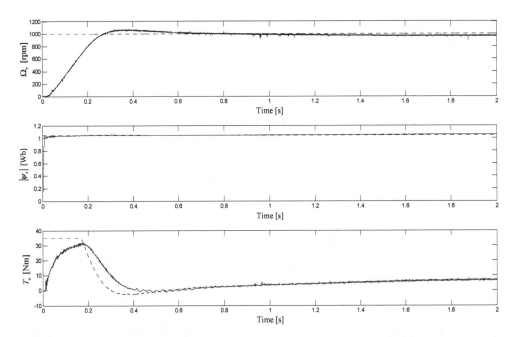

Fig. 18. Speed regulation in DTC-SVM (dotted line: speed reference; solid line: speed measured)

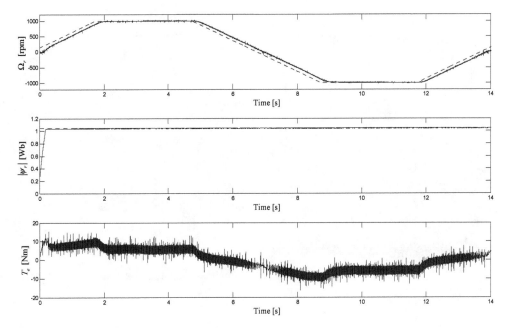

Fig. 19. Speed tracking in vector control (dotted line: speed reference; solid line: speed measured)

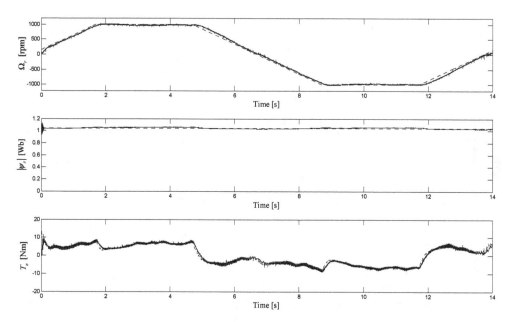

Fig. 20. Speed tracking in DCT-SVM (dotted line: speed reference; solid line: speed measured)

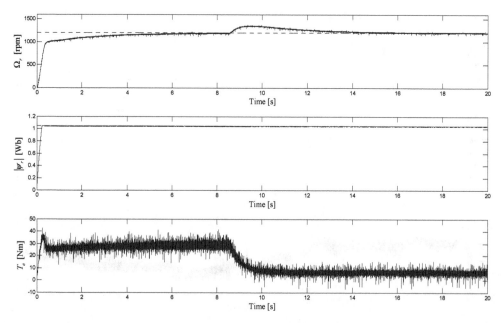

Fig. 21. Load variation in vector control (dotted line: speed reference; solid line: speed measured)

7. Educational experience

During the lectures, students become familiar with the steady-state and dynamic models of IM and the three control schemes. Then, through the experimental setup, they are made sensitive to many aspects concerning:

- modeling and simulation using the Matlab/Simulink/dSPACE environment;
- IM parameters determination;
- IM motor drives;
- some practical aspects concerning inverters and SV-PWM.

From the students' point of view, the experimental setup presented in this chapter helps them in understanding IM parameters determination, IM steady-state as well as dynamic models and IM drives. They also found that the dSPACE platform and Matlab/Simulink environment could be used for practical teaching in other courses.

From the authors' point of view, dSPACE material offers multiple advantages from the point of education and gives a powerful tool for the teaching of IM parameters identification and drives. Moreover, thanks to this experimental setup, the students can easily put into practice a lot of theoretical knowledge.

8. Conclusion

This chapter has dealt with the implementation of three control schemes of a voltage-fed inverter IM drive, namely scalar control, vector control and DTC, using a dSPACE platform and Matlab/Simulink environment. This has been successfully integrated into "Electric Drives" course which is, usually, attended by students during their fourth year of five-year electrical engineering degree at the Faculty of Engineering (FPMs) of the University of Mons (UMons) in Belgium. It helps greatly the students in understanding the theoretical concepts taught during the lectures.

Furthermore, the dSPACE platform and Matlab/Simulink environment give a powerful tool for teaching IM parameters identification and drives. The authors are planning to use more and more this experimental system in others teaching projects.

9. References

Böcker, J. & Mathapati, S. (2007). State of the art of induction motor control, *Proceedings of 2007 International Conference on Electric Machines & Drives*, ISBN 1-4244-0743-5, Antalya, Turkey, May 2007.

Bojoi, R.; Profume, F.; Griva, G.; Teodorescu, R. & Blaabjerg, F. (2002). Advanced research and education in electrical drives by using digital real-time hardware-in-the-loop simulation, *Proceedings of the 10th International Power Electronics and Motion Control Conference*, ISBN 0-7803-7089-9, Cavtat & Dubrovnik, Croatia, September 2002.

Bose, B. K. (2002). Modern Power Electronics and AC Drives, Prentice Hall PTR, ISBN 0-13-016743-6, New Jersey, USA.

Casadei, D.; Serra, G.; Tani, A. & Zarri, L. (2005). A review on the direct torque control of induction motors, *Proceedings of the 11th International Conference on Power Electronics and Applications*, ISBN 90-75815-07-7, Dresden, Germany, September 2005.

Khambadkon, A. M. & Holtz, J. (1991). Vector-controlled induction motor drive with a self-commissioning scheme, *IEEE Transactions on Industrial Electronics*, vol. 38, no. 5, (March 1991), pp. 322-327, ISSN 0278-0046.

Kishimoto, T.; Matsumoto, K.; Kamakure, T. & Daijo, M. (1986). Stability analysis of a voltage source PWM inverter-fed induction motor drive system, *Electrical Engineering in Japan*, vol. 106, no. 6, (June 1986), pp. 32-41, ISSN 0424-7760.

Luo, G.; Liu, W.; Song, K. & Zeng, Z. (2008). dSPACE based permanent magnet motor HIL simulation and test bench, *Proceedings of the 2008 IEEE International Conference on Industrial Technology*, ISBN 978-1-4244-1706-3, Chengdu, China, April 2008.

Mäki, K.; Partanen, A.; Rauhala, T.; Repo, S. & Järventausta, P. (2005). Real-time simulation environment for power system studies using RTDS and dSPACE simulators, *Proceedings of the 11th International Conference on Power Electronics and Applications*, ISBN 90-75815-07-7, Dresden, Germany, September 2005.

Mohan, N. (2001). Advanced Electric Drives: Analysis, Control and Modeling using Simulink ®, Edition MPERE, ISBN 0-9715292-0-5, Minneapolis, USA.

Padhy, P. K. & Majhi, S. (2006). Relay based PI-PD design for stable and unstable FOPDT processes, *Computer and Chemical Engineering*, vol. 30, no. 5, (April 2006), pp. 790-796, ISSN 0098-1354.

Rodriguez, J. ; Pontt, J. ; Silva, C. ; Kouro, S. & Miranda, H. (2004). A novel direct torque scheme for induction machines with space vector modulation, *Proceedings of 35th Annual IEEE Power Electronics Specialists Conference*, ISBN 0-7803-8399-0, Aachen Germany, June 2004.

Santisteban, J. A. & Stephan, R. M. (2001). Vector control methods for induction machines: an overview, *IEEE Transactions on Education*, vol. 44, no. 2, (May 2001), pp. 170-175, ISSN 0018-9359.

Trzynadlowski, A. M.; Kazmierkowski, M. P.; Graboswski, P. Z. & Bech M. M. (1999). Three examples of DSP applications in advanced induction motor drives, *Proceedings of the 1999 American Control Conference*, ISBN 0-7803-4990-3, San Diego, USA, June 1999.

Versèle, C.; Deblecker, O. & Lobry, J. (2008). Implementation of a vector control scheme using dSPACE material for teaching induction motor drives and parameters identification, *Proceedings of the 2008 International Conference on Electrical Machines*, ISBN 978-1-4244-1736-0, Vilamoura, Portugal, September 2008.

Versèle, C.; Deblecker, O. & Lobry, J. (2010). Implementation of advanced control schemes using dSPACE material for teaching induction motor drive, *International Journal of Electrical Engineering Education*, vol. 47, no. 2, (April 2010), pp. 151-167, ISSN 0020-7209.

9

Wavelet Fault Diagnosis of Induction Motor

Khalaf Salloum Gaeid and Hew Wooi Ping
University of Malaya
Malaysia

1. Introduction

The early 1980s saw the emergence of wavelets, an analysis tool that drew a lot of attention from scientists of various disciplines, mathematicians in particular, due to its promising applications.

The roots of wavelet techniques go back to 1807, when Joseph Fourier presented his theories of frequency analysis. By the 1930s, investigations were being carried out on scale-varying basis functions that would conserve energy in the computation of functional energy. Between 1960 and 1980, Guido Weiss and Ronald R. Coifman studied the reconstruction of functional elements using 'atoms'. Later, Grossman and Morlet would provide a quantum physics definition of wavelets. Stephane Mallat made an important contribution to the development of wavelets through his work in digital signal processing in 1985. The first non-trivial, continuously differentiable wavelets were created by Y.Meyer. Not long after, Ingrid Daubechies constructed a set of orthonormal basis functions that form the foundation of modern day wavelet applications (Nirmesh Yadav et al, 2004).

The present demand in the industry is for high performance electric drives that are capable of achieving speed commands accurately. Control methods have had to reach higher levels of sophistication accordingly. Induction motors, with their advantages in terms of size, cost and efficiency, are best suited to meet these growing needs (Khalaf Salloum Gaeid&Hew Wooi Ping, 2010).

In costly systems, maintenance and protection are especially essential in the prevention of system breakdowns and catastrophes. Thanks to advances in signal processing technology, it is now possible to utilize wavelet principles to efficiently diagnose and protect industrial induction motors.

Motor Current Signature Analysis (MCSA) of the stator current with wavelet to detect the fault in a broken rotor bar in the transient region was done by (Douglas et al, 2003).The analysis of the sensorless control system of induction motor with a broken rotor for diagnostics using wavelet techniques has been presented by (Bogalecka et al, 2009). (Zhang et al, 2007), used the empirical model decomposition(EMD) which deals with nonlinear systems to detect the broken rotor bar using wavelet discrete transform (WDT).(Cao Zhitong et al, 2001), used the multi resolution wavelet analysis (MRA) method to detect broken rotor bars according to the analysis of stator current. The signal was filtered, differentiated and then supplied to the Daubechies wavelet with 5 levels. (Faiz, Ebrahimi et al, 2007) presented a novel criterion to detect the broken rotor bar using time stepping finite element (TSFE) to model the broken bar faults in induction motor. (Yang et al, 2007), presented a novel method to detect the rotor broken bar using Ridge wavelet. In this paper, only one phase of the stator currents was shown to be enough to extract the characteristics of the frequency component of broken bar.(Pons-Llinares et al, 2009), presented a new method to detect a broken bar in the transient region using time motor current signature analysis (TMCSA) via frequency B-Splines. (Pineda-Sanchez et al, 2010), used fractional Fourier transform as a spectral analysis tool with the TMCSA to detect the rotor broken bar. The single mean square of discrete wavelet function computation measures whether the status of the broken rotor bar of the induction motor is healthy or faulty, using Field Programmable Gate Array(FPGA). A novelty to the weighting function was introduced by (Ordaz Moreno et al, 2008). (Abbas zadeh et al, 2001) presented a novel approach to detect the broken bar fault in squirrel cage induction motors. Two 3 HP induction motors with cast aluminum rotor bars were employed for this experiment. (Cabal-Yepez et al, 2009), used FPGA to detect a number of faults in the squirrel cage such as unbalance, faulty bearing and broken bars using parallel combination of fused fast Fourier transform(FFT) and wavelet. (J.Antonino-Daviu et al, 2009) ,presented new techniques for the detection of broken bars using high order discrete wavelet (db40) and compared it with classical methods such as the Fourier transform. To remedy the shortcomings of the FFT, (Cusido, Rosero et al, 2006) introduced spectral density on the wavelet to detect many faults in the induction motor for different load conditions (7% and 10%).A few issues were seen to feature in the use of the MCSA method for fault detection, especially when the load torque was varied. (Cusido et al, 2007& Cusido et al, 2010) presented an online system for fault detection using many wavelets like the Mexican Hat, Morlet and the Agnsis mother wavelet to detect broken bar faults.

The drawbacks of using FFT, like corrupted frequency components,the noises or other phenomena such as load torque fluctuations or supply voltage oscillations (J. Pons-Llinares et al , 2010), have been investigated by many authors for detection of broken rotor bars. The Daubechies (db) is commonly employed as the mother function to avoid low level overlapping with adjacent bands.

A 0.1 Hz resolution to detect faults in an induction motor using a combination of wavelets and power spectral density was obtained by (Hamidi et al, 2004).

The fault detection in the transient region for a broken rotor bar using the instantaneous power FFT as a medium for fault detection was presented by (Douglas, H& Pillay, P, 2005). A wavelet was used to decompose the residual stator current after filtering the noise using a Notch filter.

The wavelet indicator for detecting the broken rotor bars by calculating the absolute values of the summed coefficients in the third pattern which were normalized against the summation of the wavelet coefficient, the number of scales, and the number of samples used was presented by (Supangat et al, 2006& Supangat et al,2007).

The V/F control to detect a broken rotor bar in the induction motor was made according to the probability distribution of different operation statuses of healthy and faulty motors used by (Samsi et al, 2006). In this paper, the difference in entropy was used as a measurement indication of fault.

DWT to detect the broken rotor bar in the transient region using the slip dependant fault component according to the energy ratio of the current signal to the wavelet signal was done by Riera-Guasp et al, 2008).

The detection of the broken rotor bar fault using optimized DWT and FFT in the steady state was proposed by (J. Antonino et al, 2006). (Kia et al,2009) presented a discrete wavelet transform (DWT) for broken bar detection and diagnosis faults in induction machines in which an energy test of bandwidth with time domain analysis is performed first, after which it is applied to the stator current space vector to obtain the different broken bar fault severities and load levels.

Eigen vector as a fault indicator of stator inter turn short circuit using the Eigen vector and an energy Eigen value which contain the necessary information of the electromagnetic torque signal was presented by (Liu & Huang, 2005).

The finite element (FE) modelling of the internal faults of an induction motor. They solved the equation by the time stepping approach of a broken bar and stator shorted turns using db10 wavelet for both sinusoidal and non-sinusoidal cases has been used by (Mohammed et al, 2006& Mohammed et al, 2007).

Software diagnosis of short inter turn circuit and open circuit of the stator winding as an incipient fault was performed by (Ponci et al, 2007) to avoid any hardware cost and difficulty using wavelet decomposition for different values of stator resistance.

The MCSA technique and a wavelet to detect faults but performed the stator teeth harmonic variation using dq0 components instead of stator currents (I_{abc})was done by (Cusido et al, 2006). (Gang Niu et al, 2008) employed Bayesian belief fusion and multi agent fusion as a classifier tool to detect different faulty collected data using the signal processing techniques for smoothing and then used DWT to decompose the signals into different ranges of frequency.

Detection and diagnosis for rotor asymmetries in the induction motor based on the analysis of the stator start-up current has been done by (M. Riera-Guaspa et al, 2009). The authors extracted the harmonic component introduced by this fault. The left sideband component from the stator startup current, digital low-pass filtering (DLPF) and (DWT) are used in this technique. (C. Combastel et al, 2002) presented a comparison between model-based and signal-based approaches in the fault detection of the induction motor. The electrical variables are described according to the Park transformation model. Broken rotor and stator winding failures were investigated and the parameter variations due to heating were taken into account. (S. Radhika et al, 2010), in her MCSA-based fault diagnosis, classified WT extracted features using a Support Vector Machine (SVM). (Chen & Loparo, 1998) proposed the computation of a fault index for the stator winding faults. (Khan& Rahman, 2006) used two DWT to detect and classify the faults.

The continuous wavelet is a part of the wavelet used to detect faults especially when the overlapping between the frequency supply signal and the adjacent signal cannot be recognized. The work presented by (Ayaz et al, 2006) involved the use of six accelerometers that measured the vibration data of 5 kW and were put in independent places around the motor to detect the bearing damage.

A new technique for detecting and diagnosing faults in both stator and rotor windings using wound rotor induction motor was presented by (Saleh et al, 2005). This technique is based on a wavelet transform multi resolution analysis (WTMRA). (Cusido et al, 2007) presented both continuous and discrete wavelet to detect many mechanical and electrical induction motor faults using MCSA. (Sayed-Ahmed et al, 2007) studied the inter-turn short circuit in one phase of a stator winding of an induction motor energized from a vector controlled drive.

The induction motor requires a variable frequency three phase source for variable speed operation. One can realize this source by using a power converter system consisting of a rectifier connected to an inverter through a DC link.

In some control schemes where a three phase, variable frequency current source is required, current control loops are added to force the motor currents to follow an input reference.

Vector control techniques have been widely used for the high performance drive of induction motors. Like DC motors, torque control of induction motor is achieved by controlling the torque and flux components independently (Mohamed Boussak and Kamel Jarray, 2006). The similarities between DC and vector control are why the latter is referred to as decoupling, orthogonal or trans-vector control (Archana S. Nanoty, and A. R. Chudasama, 2008).

In this chapter, we investigate the use of the wavelet in the fault detection of vector-controlled induction motors, detection of the broken rotor bar and stator short winding faults as well as the verification of wavelet fault detection models using MATLAB.

Finally, two approaches for the protection of induction motor are examined. The first uses the Automatic Gain Control (AGC) to compensate the voltage of the induction motor to maintain satisfactory operation. The second halts operation when the fault severity becomes high. Without delving into excessively detailed results, we analyse the output relevant to wavelet detection and diagnosis.

2. Model of induction motor

The d-q dynamic model of the squirrel cage induction motor with the reference frame fixed to the stator is given by (Anjaneyulu, N .Kalaiarasi and K.S.R 2007):

$$V_s = R_s i_s + \frac{d\lambda_s}{dt} + j\omega_s M\lambda_s \tag{1}$$

$$V_r = R_r i_r + \frac{d\lambda_r}{dt} + (\omega_s - \omega_r)M\lambda_r \tag{2}$$

The electromagnetic torque is found as:

$$T_e = \frac{2pL_m}{3L_r}(i_{qs}i_{dr} - i_{ds}i_{qr}) \tag{3}$$

$$T_e = \frac{2pL_m}{3L_r}(i_{qs}\varphi_{dr} - i_{ds}\varphi_{qr}) \tag{4}$$

$$\varphi_{ds} = L_r i_{dr} + L_m i_{ds} \tag{5}$$

The field orientation is based on the following assumption:

$$\varphi_{qr} = 0 \tag{6}$$

$$\varphi_{dr} = const\,ant \tag{7}$$

The (vector) field orientation control performs the following calculations

$$\begin{bmatrix} i_{qs} \\ i_{ds} \end{bmatrix} = \begin{bmatrix} \cos\varphi_s & \sin\varphi_s \\ -\sin\varphi_s & \cos\varphi_s \end{bmatrix} \tag{8}$$

d-q to abc transformation is:

$$\begin{bmatrix} i_{as} \\ i_{bs} \\ i_{cs} \end{bmatrix} = \begin{bmatrix} 1 & 0 \\ -0.5 & -\sqrt{3}/2 \\ -0.5 & -\sqrt{3}/2 \end{bmatrix} \begin{bmatrix} i_{qs} \\ i_{ds} \end{bmatrix} \tag{9}$$

Accordingly, the rotor flux and the torque can be controlled individually through the stator current in the dq-axis so that the induction motor is transformed to a linear current and torque relationship.

3. Vector control of the induction motor

Vector control was invented by Hasse in 1969 and by Blaschke in 1972 when they demonstrated that an induction motor can be controlled like a separately excited dc motor. This brought a renaissance in the high performance control of AC drives.

Vector control of the squirrel cage induction motor is considered a fast response and high performance method to achieve variable speeds using a variable frequency source as shown in Fig.1

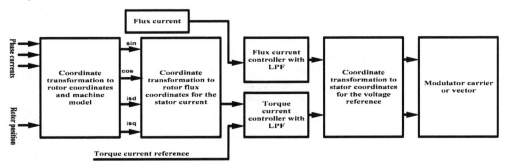

Fig. 1. Vector control principle

In the vector control method, the induction motor can be operated like a separately excited DC motor for high performance applications.

To achieve better performance, numerous speed closed loop systems have been improved. However, desired drive specifications still cannot be perfectly satisfied and/or their algorithms are too complex (J.L. Silva Neto&Hoang Le-Huy, 1996) as shown in Fig.2.

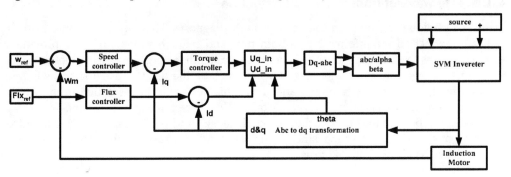

Fig. 2. Vector control implementation

Vector controlled machines need two constants as input references; the torque component (aligned with the q-axis coordinate) and the flux component (aligned with d-axis coordinate) which is simply based on projections as the control structure handles instantaneous electrical quantities. This behavior yields accurate control in both the steady state and transients and is independent of the bandwidth mathematical model of the induction motor.

Vector control is an especially advantageous solution to the problem of motor control because the torque and flux components of stator flux are easily accessible. Besides, it is free from a number of complexities that beset direct torque control (BPRA073, 1998).

There are two different strategies in vector control to obtain the rotor flux:

- Indirect control, in which the rotor flux vector is either measured by a flux sensor mounted in the air-gap or measured using the voltage equations starting from the electrical machine parameters.

- Direct control, in which the rotor flux parameter can be calculated by direct rotor speed measurement.

The condition to apply vector control to induction motors is the formulation of dynamic model equations to calculate and control the variables (AN2388, 2006) as can be seen in Fig.3, which shows the Simulink implementation of vector control using the current regulation technique. Note that both the magnitude and phase alignment of vector variables is controlled.

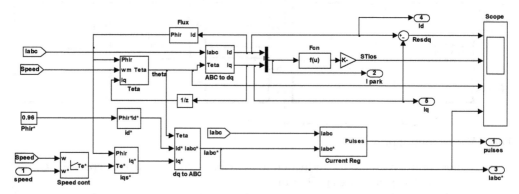

Fig. 3. Vector control implemented using Simulink.

4. Discrete wavelet transform fault detection

Wavelet techniques are new in the field of fault diagnosis. They are useful due to their ability to extract all the information in both time and frequency domain. They provide a sensitive means to diagnose the faults in comparison to other signal processing methods like the Fourier Transform, the drawbacks of which include the need to use a single window function in all frequency components and the acquisition of linear resolution in the whole frequency domain. This is an important reason for the interest in wavelets in time–frequency analysis as can be seen in (M. Riera-Guaspa et al, 2009). (Andrew K.S. et al, 2006) presented a review of the diagnosis of machines using the condition-based maintenance approach. There are two levels of fault diagnosis:

1. Traditional control
2. Knowledge based fault diagnosis

Fault diagnosis techniques contain the feature extraction module (wavelet), feature cluster module and the fault decision module (1). Indicators of faults include the negative sequence current, impedance and the Park's vector.

Motor Current Signature Analysis (MCSA) is used to diagnose the stator short circuit fault. Multi resolution analysis and good time localization are particularly useful characteristics of wavelets in the context of fault diagnosis.

Signal processing techniques like the FFT are based on the assumptions of constant stator fundamental frequency, load, motor speed and the assumption that the load is sufficient. Wavelet transformation is of many kinds but in this chapter the authors will introduce the most important among them:

1. Discrete wavelet transformation
2. Continuous wavelet transformation
3. Wavelet packet decomposition transformation

The wavelet is divided into two main groups. One is the discrete wavelet transform represented in the following Eq.:

$$DWT(m,k) = \frac{1}{\sqrt{a_0^m}} \Sigma x(n) g(\frac{k - nb_0 a_0^m}{a_0^m}) \tag{10}$$

Where $g(n)$ is the mother wavelet $x(n)$ is the input signal and the scaling and translation parameters " a "and " b " are functions of the integer parameter m (M. Sushama et al, 2009).

The second wavelet type is the continuous wavelet transform (CWT) which can be represented as follows:

$$\omega(m,n) = \int_{-\infty}^{\infty} f(t) \psi_{m,n}^{*}(t) dt \tag{11}$$

* denotes the complex conjugate, where f (t) is the waveform signal and ψ (t) is a wavelet.

$$\psi_{m,n}(t) = 2^{-1/2} \psi(2^{-m} t - n) \tag{12}$$

Where m and n are the wavelet dilation and translation used to transform the original signal to a new one with smaller scales according to the high frequency components. This relation is valid for the orthogonal basis of wavelet transform (a =2 and b =1). In the following continuous wavelet transform, a is the scale parameter, b is the time parameter.

$$\omega_{a,b}(t) = |a|^{-1/2} \psi(\frac{t-b}{a}) \tag{13}$$

CWT is divided into the Real wavelet as can be seen in table 1 and the Complex wavelet as in table 2.

In digital computers, the discrete wavelet transform is a good choice. The mother wavelet is scaled to the power of 2 (R. Salehi Arashloo&A. Jalilian, 2010).

The continuous wavelet transform (CWT) was developed as an alternative approach to overcome the resolution problem as is shown in Table 1 (Lorand SZABO et al, 2005).

Beta Wavelet	$\psi_{beta}(t/\alpha,\beta) = (-1)dp(t/\alpha,\beta)/dt$
Hermitian Wavelet	$\psi_n(t) = (2n)^{-n/2}C_n H_n(t/\sqrt{2})e^{(-1/2n)t^2}$
Mex.hat wavelet	$\psi(t) = (2/\sqrt{3}\sigma\pi^{1/4})(1-t^2/\sigma^2)e^{(-t^2/2\sigma^2)}$
Shannon wavelet	$\psi(t) = 2\sin c(2t) - \sin c(t)$

Table 1. Continuous real wavelet transform

Mexican hat wavelet	$\psi(t) = \dfrac{2}{\sqrt{3}}\pi^{-0.25}\left(\sqrt{\pi}(-t^2)e^{-0.5t^2} - \sqrt{2}it\sqrt{\pi}erf\left(\dfrac{i}{\sqrt{2}}t\right)(1-t^2)e^{-0.5t^2}\right)$
Morlet wavelet	$\psi(t) = (C\pi^{-(1/4)})e^{-1/2t^2}(e^{it}-k)$
Shannon wavelet	$\psi(t) = \sin c(t)e^{-j\pi t}$
Modified Morlet wavelet	$\psi(t) = C_\psi \cos(\omega_o t)\sec h(t)$

Table 2. Continuous complex wavelet transform

Two properties of the wavelet are noteworthy:
1. When a wavelet satisfies an admissibility condition, a signal with finite energy can be reconstructed without needing all values of its decomposition. The admissibility condition is represented by the following equation:

$$\int\frac{|\psi(\omega)|^2}{|\omega|}d\omega < +\infty \tag{14}$$

Where ψ (ω) is the Fourier transform of the wavelet function ψ (t) used to investigate the signals and then to reconstruct them without losing any information. According to the admissibility condition, the Fourier transform goes to zero as is shown in the following equation:

$$|\psi(\omega)|^2 = 0 \tag{15}$$

Another important property of the wavelet is:

$$\int\psi(\omega) = 0 \tag{16}$$

2. To remedy the squared relationship between the time bandwidth product of the wavelet transform and the input signal, certain regularity conditions are imposed so as to ensure the smoothness and concentration of the wavelet function in both time and frequency domains.

The decomposition can be implemented using filtering and down-sampling, and can be iterated, with successive approximation as in (Turkmenoglu, 2010).

The total decomposition levels (L) can be calculated according to the following relationship:

$$L \geq \frac{\log(\frac{fs}{f})}{\log(2)} + 1 \tag{17}$$

These bands can't be changed unless a new acquisition with different sampling frequency is made, which complicate any fault detection based on DWT, particularly in time-varying conditions (Yasser Gritli et al, 2011).

When (18) is applied at a sampling frequency of 1 kHz, a six level decomposition occurs. Table 3 depicts the frequency bands for each wavelet signal.

$$L = \frac{\log(\frac{1000}{50})}{\log(2)} + 1 = 6 \text{ levels} \tag{18}$$

Approximations «aj»	Frequency bands (Hz)	Details «dj»	Frequency bands (Hz)
a_6	[0-16.125]	d_6	[16.125-32.25]
a_5	[0-32.25]	d_5	[32.250-64.5]
a_4	[0 - 64.50]	d_4	[64.50-125.0]
a_3	[0 – 125.0]	d_3	[125.0-250.0]
a_2	[0- 250.0]	d_2	[250.0-500.0]
a_1	[0-500.0]	d_1	[500.0-1000.0]

Table 3. Frequency bands for the six levels of wavelet signals

The data required to analyze the signal depends on both the sampling frequency ( ) and the resolution (R) as in (19):

$$D_{required} = f_s / R \tag{19}$$

The interpretation of the above table is shown in Fig 4:

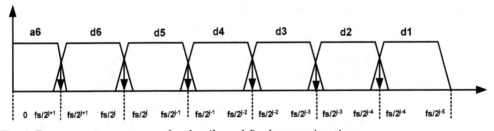

Fig. 4. Frequency range covers for details and final approximation

The MATLAB signal processing toolbox software provides the Filter Design and Analysis tool (FDATool) that enables the design of a low pass filter and high pass filter and then export the filters coefficients to a matching filter implemented as can be seen in the Fig.5.

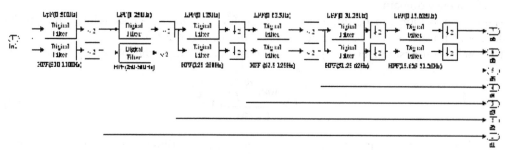

Fig. 5. Wavelet decomposition levels using FDA matlab tool box

An alternative way to performing the same task is through the DWT dyadic filter from MATLAB/Simulink. Although for this method, the wavelet coefficients may need to be calculated using the following MATLAB instruction:

$$[Lo\_D, Hi\_D, Lo\_R, Hi\_R] = wfilters \ ('db10') \qquad (20)$$

Where Lo_D,Hi_D,Lo_R,Hi_R represent low pass filter decomposition,high pass filter decomposition ,low pass filter reconstruction and high pass filter reconstruction respectively. This is shown in Fig.6.

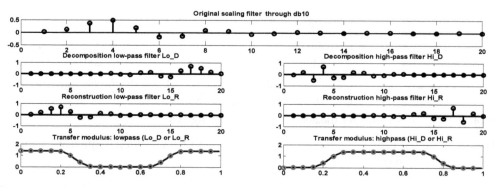

Fig. 6. Wavelet decomposition and reconsstruction f filters

The decomposition of the wavelets was impemented using the above relation as can be seen in Fig.7.

The above two circuits are exactly the same. A wavelet-transform-based method was developed for diagnosis and protection the induction motor against broken rotor bar and short stator windings. Detailed information is obtained from the high pass filters and the approximation information is obtained from low pass filter. Daubechies wavelet (db10) is used to analayze stator current as in Fig8.

The construction of DWT is followed by implementing the criterion of fault detection of induction motor faults. The criteria used to detect the induction motor faults depend on the relationship between maximum detail energy (d6) and the original stator current (Ia) as is shown in Fig.9.

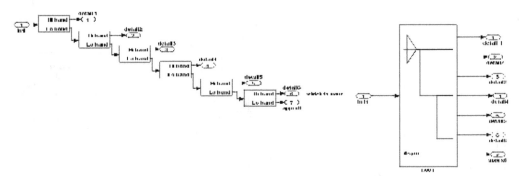

Fig. 7. DWT filters and levels using simulink

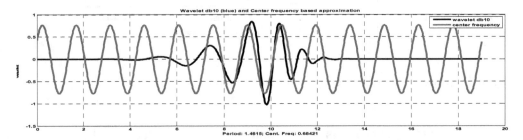

Fig. 8. Daubechies wavelet (db10) and central frequency

Offline calculations of maximum energy were done using MATLAB instructions as in the following wavelet program:

load Ia

[C, L] = wavedec (Ia, 6,'db10');

[Ea, Ed] = wenergy(C, L)

Ea is the percentage of energy corresponding to the approximation, Ed is the vector containing the percentages of energy corresponding to the details, C is the wavelet decomposition vector and L is the bookkeeping vector.

Ea = 99.5370(a6)

Ed =0.0000(d1)   0.0000(d2)   0.0000(d3)   0.0001(d4)   0.0010(d5)   0.4619(d6)

The Wavelet coefficients, the energy of the details of any signal at level j can be expressed as (M. Sabarimalai Manikandan, and S. Dandapat, 2007):

$$E_j = \sum d_{j,k}^{\,2} \tag{21}$$

$$d_{j,k} = \langle x(t), \psi_{j,k} \rangle = \frac{1}{\sqrt{2^j}} \int x(t)\psi(2^j t - k)dt \tag{22}$$

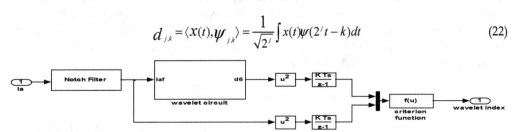

Fig. 9. Wavelet index unit with conditioning signal

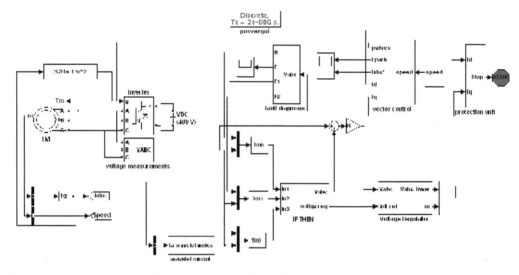

Fig. 10. Proposed circuit of induction motor fault diagnosis with wavelet

M. specifications	Unit	Value
power	w	370
Current	ampere	1.7
Voltage (delta)	volt	230
Rated speed	RPM	2800
No. of pole		2
Moment of inertia	Kgm²	3.5e-4
Stator resist.	ohm	24.6
Rotor resist.	ohm	16.1
Stator induct.	henry	40e-3
Rotor induct.	henry	40e-3

Table 4. Induction motor specifications

5. Fault detection

Electric drives are used in safety-critical applications or industrial processes where the immense costs of unplanned stops are unacceptable. Fault detection depends on the availability of information from the system. In this work, the fault detection is done using wavelet for analysis of stator current as can be shown in Fig.11 for the healthy case, Fig.12 for broken rotor bar case and Fig.13 for the stator short winding case respectively.
The wavelet criterion of fault detection is:

$$W_{indx} = abs(energy(d6)) / average(energy(Ia)) \qquad (23)$$

In MATLAB/Simulink, an assessment of operating features of the proposed scheme is performed. Two faults are investigated: stator short winding and broken rotor bar.

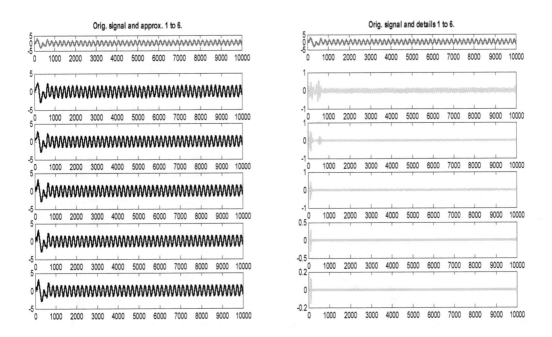

Fig. 11. Approximation and details signal in healthy motor

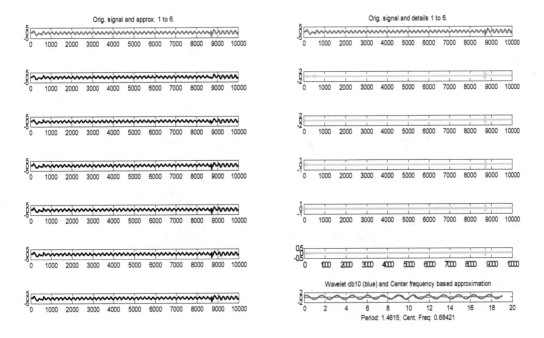

Fig. 12. Approximation and details signal in one broken rotor bar fault

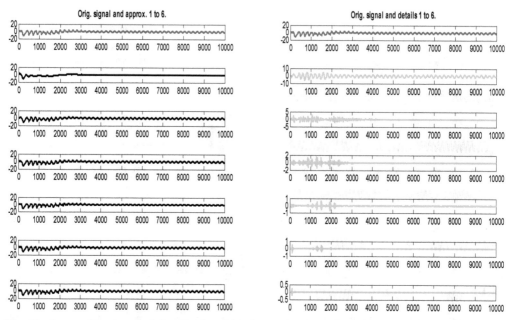

Fig. 13. Approximation and details signal in short stator winding fault

This wavelet is also used as fault indicator or wave index as is shown in Fig14.

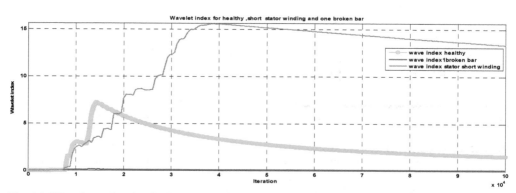

Fig. 14. Wavelet index for fault detection

5.1 Broken rotor bar
Key reasons for a broken rotor bar are (Ahmed Y. Ben Sasi et al, 2006):
1. Direct on line starting which leads to excessive heating and mechanical problems.
2. Variable mechanical load.
3. Unsatisfactory rotor cage manufacturing.

Broken rotor bar faults can be simulated by connecting three resistances with the rotor resistance so that by increasing one of the rotor phase resistances, the broken rotor bar equivalent resistance can be computed as in (24).

$$R_{brk} \cong (0.33/4)R_r z_{nb} / N\char`\^2_s \tag{24}$$

The external added resistances are changed in 0.0833 Ω steps, which represents the difference between the reference rotor resistance and the original rotor resistance for one broken rotor bar .Reference rotor resistance depends on the number of broken bars and the total number of rotor bars (Hakan Calıs& Abdulkadir Cakır, 2007). The resistance of induction motor rotor bar is assumed to be high.

(Levent Eren, & Michael J. Devaney, 2004), presented the bearing fault defects of the induction motor WPT decomposition of 1 Hp induction motor stator current through the test of RMS for both healthy and faulty bearings.

5.2 Stator shorting the winding

More than 30% of all motor faults are caused by failure of the motor winding due to insulation problems. For the stator short circuit winding fault, the stator resistance of the induction motor is connected to the parallel variable resistance which is reduced according to the following formula:

$$R_{sh} = 0.1R_{org} \qquad (25)$$

The majority of induction motor winding failures proceed gradually from lower short circuit current to a higher level and finally break as can be seen in (Dimas et al, 2010).

To check the validity of the wavelet fault detection of both stator winding and broken rotor bar units as well as when the motor is in a healthy condition,MATLAB/Simulink's Predicted Model Block (PED) is used to verifiy the wavelet detection units as is shown in subsequent figures.

In the healthy induction motor, transfer finction of the wavelet unit (interval test) is:

$$G_{healthy} = \frac{-1.3e-166z^3 + 8.2e-167z^2}{z^4 - 0.5z^3 - 0.5z^2 + 7.7067e-8z - 1.1592e-8} \qquad (26)$$

And the noise model of unit is:

$$G_{Noise} = \frac{z^4 + 0.0027766z^3 + 0.00025661z^2}{z^4 - 2.3152z^3 + 1.2203z^2 + 0.5013z - 0.40632} \qquad (27)$$

The predicted and noise model of the wavelet detection unit in the healthy case is shown in Fig15.

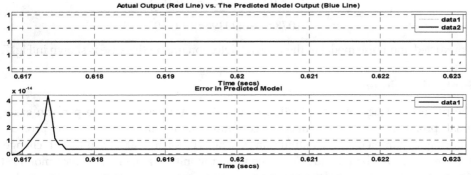

Fig. 15. Predicted model output and its noise model for wavelet detcteion unit in the lealthy case

In the stator short winding fault, transfer function of the wavelet unit (interval test) is:

$$G_{statorshort} = \frac{1.1665e-158z^3 - 2.2477e-158z^2}{z^4 - 0.5z^3 - 0.5z^2 + 2.9406e-10z - 3.5998e-11}$$ (28)

The noise model of the unit is:

$$G_{noise\,model} = \frac{z^4 + 0.0020325z^3 - 0.00119535z^2}{z^4 - 2.4623z^3 + 1.4434z^2 + 0.5z - 0.48113}$$ (29)

The predicted and noise model of the wavelet detection unit in this case is shown in Fig16.

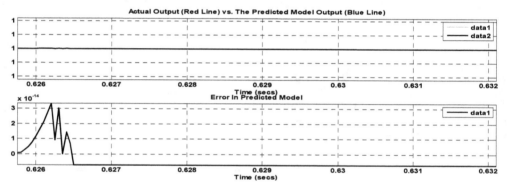

Fig. 16. predicted model output and its noise model for wavelet detcteion unit in the stator short winding case

In the broken rotor bar fault, transfer function of the wavelet unit (interval test) is:

$$G_{brokenbar} = \frac{1.17382e-159z^3 - 2.0472e-159z^2}{z^4 - 0.5z^3 - 0.5z^2 - 3.4564e-11z - 4.1663e-11}$$ (30)

The noise model of the unit is:

$$G_{noise\,model} = \frac{z^4 + 0.0035537z^3 - 0.0011322z^2}{z^4 - 2.1078z^3 + 0.90396z^2 + 0.503z - 0.30002}$$ (31)

The predicted and noise model of the wavelet detection unit in this case is shown in Fig17.

6. Protection circuit

In the protection stage of the induction motor, there are many steps to perform exact or optimal protection of the circuit like: condition monitoring which is the process of monitoring a parameter of condition in machinery, such that a significant change is indicative of a developing failure. Many condition-monitoring methods, which monitor the motor's condition using only the currents and voltages of the motor, are preferred due to their low cost and non-intrusiveness (Zhang et al, 2011). For reliable operation of ajustable

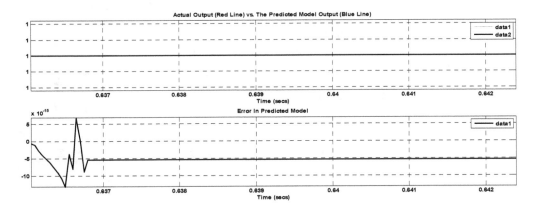

Fig. 17. Predicted model output and its noise model for wavelet detcteion unit in the broken rotor.

speed drive systems, the vulnerable components of the power converter, cable, and motor must be monitored, since failure of a single component can result in a forced outage of the entire system (Lee, et al, 2011).

In this chapter, two approaches are used to treat the faults mentioned above. First, voltage regulation with automatic gain control (AGC) is used to control the voltage after the occurrence of fault and hence the speed to maintain the operation of the induction motor as in (32):

$$AGC = reciprocal(mean(real(f_{vd})^2 + real(f_{vq})^2) \tag{32}$$

This is shown in Fig18:

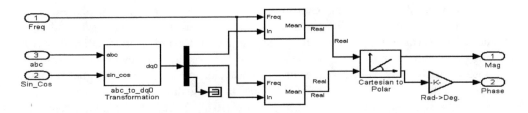

Fig. 18. Automatic gain controller circuit

The fault diagnosis condition depends also upon an optimization technique of induction motor flux as in (33).

$$\varphi_{optimal} = \mid T_e * \sqrt{R_r \Big/ (3/2*p)} \mid \tag{33}$$

The proposed circuit of the wavelet fault diagnosis is shown in Fig10. The last stage of the protection is to stop the motor operation when the fault severity becomes high and cannot be controlled according to the following criteria.

$$if \; \max(Iq) > 180 \; \& \, bad \; flux \, due \, to \, bad \, torque \;\; then \; stop \, the \; operation \qquad (34)$$

The protection circuit as in the Fig.19

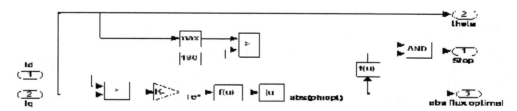

Fig. 19. Protection and checking of the optimal flux unit

Since this work focuses on the wavelet, the results of this stage are not included here. They are expected to be presented by the authors in a paper dealing with the protection of induction motors.

7. Simulation results

Computer simulations using Matlab/Simulink have been performed for assessment of operating features of the proposed scheme. The simulation involved a startup of an induction motor 0.5Hp, 230V, 50 Hz as is shown in Table 4. All pertinent mathematical models have been developed individually, using Simulink blocks for the power, electronic converters and the motor. The developed torque increases rapidly due to the slip speed, which is maximum at starting and decreases as the speed of the motor is increased. The classic vector control scheme consists of the torque and flux control loops. For the vector control of induction motors, the rotor field orientation has the advantage of easy decoupling of the torque and flux components of the controlled variable. The speed loop utilizes a PI controller to produce the quadrature axis currents which serves as the torque command (K_p =18 and K_i=5). Better PI tuning results in a better wavelet waveform. One phase of stator current is used as the input to the wavelet circuit. It is clear from Fig.11, Fig.12 and Fig.13 that detailed information is extracted from the high pass filter and approximation data from the low pass filter to show the exact location and time of the fault even for the transient response of the healthy case. A higher order of wavelet introduces better signal accuracy. Fig.14 shows the wavelet index with the threshold of the healthy case between 2.5 and 7. In this work, the stator short winding was high index (16.6), the range of this fault being 7-17 due to the severity of fault; the broken rotor bar was very light (0.12) and the range of this fault was between 0 and 0.2.The figure did not start from zero because there is a 1 sec delay for protection. Careful tuning of AGC is needed for the PID controller to get best results, in this work (K_i=100, K_p = 1 and K_d=0).

Fig.15, Fig.16 and Fig.17 show the outputs of the PEM (Predictive Error Method) estimator block which estimates linear input-output polynomial models in Simulink for the healthy case, stator short winding and broken rotor bar respectively. The monitoring of the optimal flux is another fault diagnosis indicator in this work which depends on both torque and rotor resistance as well as on the poles of the induction motor.

The last stage of the operation is the protection. Fig.19 shows the optimal flux and protection mechanism which depends on both torque, flux and on the phase of I_q. If this phase is more than 180, this will lead to bad torque and flux and will indicate the need to stop the operation of induction motor.

8. Conclusion

The scalar control of induction motor drives have many drawbacks including slow response, unsuitable performance, torque ripple and impossibility of operation at all points of speed torque curve, vector control is implemented on an induction motor drive to solve these imperfections.

Three phase sinusoidal voltage is converted to dc voltage with the help of universal diode bridge rectifier. Current source inverters which dominate modern adjustable speed AC drives are not free from certain disadvantages.

In particular the high switching rate needed for good quality of the current fed to the motor cause's losses, electromagnetic interference etc.Accords to the results, the algorithm used is very effective and have been succeeded in maintain both speed and torque.

There are many conclusions can be included due to this work:

- The wavelet is considered as powerful tools in the fault detection and diagnosis of induction motors.
- Many wavelet classes can be generated by different kinds of mother wavelets and can be constructed by filters banks.
- The improvement of fault detection and diagnosis can be exploiting the wavelet properties to get high detection and diagnostics effectiveness.
- Theories of wavelet need to be pushed forward to insure best choosing of mother wavelet.
- The wavelet index can distinguish correctly between the faults and healthy induction motor.
- Matlab/Simulink axcellent package for both simulations and practicle experiments in the diagnostic of induction machines with wavelet.

9. References

Abbas zadeh, K., Milimonfared, J., Haji, M., & Toliyat, H. (2001). Broken bar detection in induction motor via wavelet transformation. *The 27th Annual Conference of the IEEE Industrial Electronics Society*, Vol.1, pp. 95 –99.

Ahmed Y. Ben Sasi, Fengshou Gu, Yuhua Li and Andrew D. Ball (2006). A validated model for the prediction of rotor bar failure in squirrel-cage motors using instantaneous angular speed. *Mechanical Systems and Signal Processing*, Vol.20, No 7, pp. 1572-1589.

Hakan Calıs& Abdulkadir Cakır (2007). Rotor bar fault diagnosis in three phase induction motors by monitoring fluctuations of motor current zero crossing instants. *Electric Power Systems Research*, Vol.77, pp. 385–392.

Andrew K.S. Jardine, Daming Lin, Dragan Banjevic (2006). A review on machinery diagnostics and prognostics implementing condition-based maintenance. *Mechanical Systems and Signal Processing*, Vol.20, pp.1483–1510.

Anjaneyulu, N .Kalaiarasi and K.S.R (2007). Adaptive Vector Control of Induction Motor Drives. *International Journal of Electrical and Engineering*, Vol.1 .No.2, 239-245.

Archana S. Nanoty, and A. R. Chudasama (2008). Vector Control of Multimotor Drive. *Proceedings of World Academy of Science, Engineering and Technology,* Vol.35, pp. 2070-3740.

Ayaz, E., Ozturk, A., & Seker, S. (2006). Continuous Wavelet Transform for Bearing Damage Detection in Electric Motors. *IEEE Electrotechnical Conference, MELECON.* Mediterranean, pp. 1130 –1133.

Bogalecka, Piotr Kołodziejek and Elz bieta (2009). Broken rotor bar impact on sensorless control of induction machine. *The International Journal for Computation and Mathematics in Electrical and Electronic Engineering,* Vol.28, No.3, pp.540-555.

BPRA073. (1998). Field Orientated Control of 3-Phase AC-Motors. *Texas Instruments Europe.*

C. Combastel, S. Lesecq, S. Petropol, S. Gentil (2002). Model-based and wavelet approaches to induction motor on-line fault detection. *Control Engineering Practice,* Vol.10, No.5, pp.493–509.

Cabal-Yepez, E., Osornio-Rios, R., Romero-Troncoso, R., Razo-Hernandez, J., & Lopez-Garcia, R. (2009). FPGA-Based Online Induction Motor Multiple-Fault Detection with Fused FFT andWavelet Analysis. *International Conference on Reconfigurable Computing and FPGAs,* pp. 101 –106.

Cao Zhitong, Chen Hongping, He Guoguang, Ritchie, E (2001). Rotor fault diagnosis of induction motor based on wavelet reconstruction"*Proceedings of the Fifth International Conference on Electrical Machines and Systems,* ICEMS, Vol.1, pp. 374 – 377.

Chen, C.-M., & Loparo, K. (1998). Electric fault detection for vector-controlled induction motors using the discrete wavelet transform. *Proceedings of the American Control Conference,* Vol.6, pp. 3297 –3301.

Cusido, J., Rosero, J., Cusido, M., Garcia, A., Ortega, J. L., & Author, Q. (2007). On-Line System for Fault Detection in Induction Machines based on Wavelet Convolution. *IEEE Conf.On Power Electronics Specialists, PESC,* pp. 927-932.

Cusido, J., Rosero, J., Cusido, M., Garcia, A., Ortega, J., & Romeral, L. (2007). On-Line System for Fault Detection in Induction Machines Based on Wavelet Convolution. *IEEE Instrumentation and Measurement Technology Conference Proceedings, IMTC,* pp. 1 - 5.

Cusido, J., Rosero, J., Ortega, J., Garcia, A., & Romeral, L. (2006). Induction Motor Fault Detection by using Wavelet decomposition on dq0 components. *IEEE International Symposium on Industrial Electronics,* Vol.3, pp. 2406 -2411.

Cusido, J., Rosero, J., Romeral, L., Ortega, J., & Garcia, A. (2006). Fault Detection in Induction Machines by Using Power Spectral Density on the Wavelet Decompositions. *37th IEEE Conference on Power Electronics Specialists PESC,*pp. 1-6.

Dimas Anton A, Syafaruddin, Dicky N Wardana, M.H.Purnomo, Takashi Hiyama (2010). Characterization of Temporary Short Circuit in Induction Motor Winding using Wavelet Analysis. *Proceedings of the International Conference on Modelling, Identification and Control,* pp. 477 -482.

Douglas, H., & Pillay, P. (2005). The impact of wavelet selection on transient motor current signature analysis. *IEEE International Conference on Electric Machines and Drives,* pp. 80 – 85.

Douglas, H., Pillay, P., & Ziarani, A. (2003). Detection of broken rotor bars in induction motors using wavelet analysis. *International IEEE Conf. On Electric Machines and Drive, IEMDC*, Vol.2, pp. 923 –928.

Levent Eren and Michael J. Devaney (2004). Bearing Damage Detection via Wavelet Packet Decomposition of the Stator Current. *IEEE Transactions on Instrumentation and Measurement*, Vol.53, No.2, pp. 431-436.

Faiz, J., Ebrahimi, B., Asaie, B., Rajabioun, R., & Toliyat, H. (2007). A criterion function for broken bar fault diagnosis in induction motor under load variation using wavelet transform. *International Conference on Electrical Machines and Systems, ICEMS*, pp. 1249 –1254.

Gang Niu, Achmad Widodo, Jong-Duk Son, Bo-Suk Yang, Don-Ha Hwang , Dong, Sik Kang (2008). Decision level fusion based on wavelet decomposition for induction motor fault diagnosis using transient current signal. *Expert Systems with Applications*, Vol.35, pp.918–928.

Hamidi, H., Nasiri, A., & Nasiri, F. (2004). Detection and isolation of mixed eccentricity in three phase induction motor via wavelet packet decomposition. *5th Asian Control Conference*, Vol.2, pp. 1371 –1376.

J. Antonino-Daviu, M. Riera-Guasp, J. Roger-Folch, F.Martinez-Gimenez, A. Peris (2006). Application and optimization ofthe discrete wavelet transform for the detection of broken rotor bars in induction machines. *Applied and Computational Harmonic Analysis*, Vol.21, No.2, pp.268–279.

J. Cusido,L. Romerala, J.A. Ortega, A. Garcia and J.R. Riba. (2010). Wavelet and PDD as fault detection techniques. *Electric Power Systems Resaerch*, Vol.80, No.8, pp 915-924.

J.Antonino-Daviu, P.Jover Rodriguez, M.Riera-Guasp M. Pineda-Sanchez, A.Arkkio (2009). Detection combined faults in induction machines with stator parallel branches through the DWT of the startup current. *Mechanical Systems and Signal Processing*, Vo.23, pp.2336–2351.

J.L. Silva Neto, Hoang Le-Huy (1996). Fuzzy Logic Based Controller for Induction Motor Drives. *IEEE, IECON*, pp. 631-634.

Mohamed Boussak and Kamel Jarray (2006). A High Performance Sensorless Indirect Stator Flux Orientation Control of Induction Motor Drive. *IEEE trans. Industrial Electronics*, Vol. 53, pp. 41-49.

Nirmesh Yadav, Sharad Sharma, Vikram M. Gadre, &ldquo (2004). Wavelet Analysis and Applications. *New Age International Publishers*, ISBN 81-224-1515-6.

Khalaf Salloum Gaeid, Hew Wooi ping(2010). Diagnosis and Fault Tolerant Control of the Induction Motors Techniques: A review. *Australian Journal of Basic and Applied Sciences*, Vol.4 No.2, pp. 227-246.

Khan, M., & Rahman, M. (2006). Discrete Wavelet Transform Based Detection of Disturbances in Induction Motors. *Electrical and Computer Engineering, ICECE*, pp. 462 –465.

Kia, S., Henao, H., & Capolino, G.-A. (2009). Diagnosis of Broken-Bar Fault in Induction Machines Using Discrete Wavelet Transform without Slip Estimation. *IEEE Transactions on Industry Applications*, Vol.45, No.4, pp1395 –1404.

Lee, S. B.; Yang, J.; Hong, J.; Yoo, J.-Y.; Kim, B.; Lee, K.; Yun, J.; Kim, M.; Lee, K.-W.; Wiedenbrug, E. J.; Nandi, S. (2011). A New Strategy for Condition Monitoring of Adjustable Speed Induction Machine Drive Systems. *IEEE Transactions on Power Electronics*, Vol.26, No.2, pp.389 - 398.

Liu, T., & Huang, J. (2005). A novel method for induction motors stator interturn short circuit fault diagnosis by Systems, Proceedings *of the Eighth International Conference on Electrical Machines and Systems, ICEMS*, Vol. 3, pp. 2254 – 2258.

Lorand SZABO, JenoBarna DOBAI, Karoly Agoston BIRO (2005). Discrete Wavelet Transform Based Rotor Faults Detection Method for Induction Machines. *Intelligent Systems at the Service of Mankind*, Vol. 2, pp. 63-74.

M. Riera-Guaspa, J. Antonino-Daviua,J. Rusekb, J. Roger Folch (2009). Diagnosis of rotor asymmetries in induction motors based on the transient extraction of fault components using filtering techniques. *Electric Power Systems Research*, Vol.79, No.8, pp.1181–1191.

M. Sabarimalai Manikandan, and S. Dandapat (2007). Wavelet energy based diagnostic distortion measure for ECG. *Biomedical Signal Processing and Control*, Vol.2, No.2, pp. 80-96.

M. Sushama, G. Tulasi Ram Das and A. Jaya Laxmi (2009). Detection of High-Impedance Faults in Transmission. ARPN *Journal of Engineering and Applied Sciences*, Vol.4, No. 3, pp.6-12.

Mohammed, O., Abed, N., & Ganu, S. (2006). Modelling and Characterization of Induction Motor Internal Faults Using Finite-Element and Discrete Wavelet Transforms. *IEEE Transactions on Magnetics*, Vol.42, No.10, pp.3434 –3436.

Mohammed, O., Abed, N., & Garni, S. (2007). Modeling and characterization of induction motor internal faults using finite element and discrete wavelet transforms. *IEEE Symposium on Electric Ship Technologies, ESTS*, pp. 250 – 253.

AN2388 Application Note, (2006). Sensor Field Oriented Control (IFOC) of Three-Phase AC Induction Motors Using ST10F276.

Ordaz Moreno, A. de Jesus Romero-Troncoso, R. Vite-Frias (2008). Automatic Online Diagnosis Algorithm for Broken-Bar Detection on Induction Motors Based on Discrete WaveletTransform for FPGA Implementation. *IEEE Transactions on Industrial Electronics*, Nol.55, No.5, pp2193 –2202.

Pineda-Sanchez, M., Riera-Guasp, M., Antonino-Daviu, J. A., Roger-Folch, J., Perez-Cruz, J., & Puche-Panadero, R. (2010). Diagnosis of Induction Motor Faults in the Fractional Fourier Domain. *IEEE Transactions on Instrumentation and Measurement*, Vol.59, No 8, pp.2065 - 2075.

Ponci, F., Monti, A., Cristaldi, L., & Lazzaroni, M. (2007). Diagnostic of a Faulty Induction Motor Drive via Wavelet Decomposition. *IEEE Transactions on Instrumentation and Measurement*, Vol. 56 No.6, pp. 2606 –2615.

Pons-Llinares, J., Antonino-Daviu, J., Riera-Guasp, M.-S. M., & Climente-Alarcon, V. (2009). Induction motor fault diagnosis based on analytic wavelet transform via Frequency B-Splines. *IEEE International Symposium on Diagnostics for Electric Machines, Power Electronics and Drives, SDEMPED*, pp. 1-7.

R. Salehi Arashloo, A. Jalilian (2010). Design, Implementation and Comparison of Two Wavelet Based Methods for the Detectionof Broken Rotor Bars in Three Phase Induction Motors. *1st Power Electronic & Drive Systems & Technologies Conference, PEDSTC*, pp. 345 –350.

Riera-Guasp, M., Antonino-Daviu, J., Pineda-Sanchez, M., Puche-Panadero, R., & Perez-Cruz, J. (2008). A General Approach for the Transient Detection of Slip-Dependent Fault Components Based on the Discrete Wavelet Transform. *IEEE Transactions on Industrial Electronics*, Vol.55, No.12, pp.4167 –4180.

S. Radhika, G.R. Sabareesh, G. Jagadanand, V. Sugumaran (2010). Precise wavelet for current signature in 3φIM. *Expert Systems with applications*, Vol.37, No.1, pp.450–455.

Saleh, S., Khan, M., & Rahman, M. (2005). Application of a wavelet-based MRA for diagnosing disturbances in a three phase induction motor. *5th IEEE Conf. on Diagnostics for Electric Machines, Power Electronics and Drives, SDEMPED*, pp. 1-6.

Samsi, R., Rajagopalan, V., & Ray, A. (2006). Wavelet-based symbolic analysis for detection of broken rotor bars in inverter-fed induction motors. *American Control Conference*, pp. 3032-3038.

Sayed-Ahmed, A., Sizov, G., & Demerdash, N. (2007). Diagnosis of Inter-Turn Short Circuit for a Polyphase Induction Motor in Closed-Loop Vector-Controlled Drives. *IEEE ,42nd IAS Annual Meeting*, pp. 2262 – 2268.

Supangat, R., Ertugrul, N. W., & Grieger, J. (2006). Detection of broken rotor bars in induction motor using starting-current analysis and effects of loading. *IEEE Proceedings on Electric Power Applications*, Vol.153:6, pp. 848 –855.

Supangat, R., Grieger, J., Ertugrul, N., Soong, W., Gray, D., & Hansen, C. (2007). Detection of Broken Rotor Bar Faults and Effects of Loading in Induction Motors during Rundown. *IEEE International Conference on Electric Machines & Drives, IEMDC*, Vol.1, pp. 196 – 201.

Turkmenoglu, M. A. (2010). Wavelet-based switching faults detection in. *IET Science, Measurement and Technology*, Vol.4, No.6, pp.303–310.

Yang, C., Cui, G., Wei, Y., & Wang, Y. (2007). Fault Diagnosis for Induction Motors Using the Wavelet Ridge. *Second International Conference on Bio-Inspired Computing: Theories and Applications, BIC-TA*, pp. 231 – 235.

Yasser Gritli, Andrea Stefani,Claudio Rossi, Fiorenzo Filippetti and Abderrazak Chatt (2011). Experimental validation of doubly fed induction machine electrical faults diagnosis under time-varying conditions. *Electric Power Systems Research*, Vol.81, No.3, pp.751-766.

Zhang Jian wen, Zhu Ning-hui, Yang Li, Yao Qi, Lu Qing (2007). A Fault Diagnosis Approach for Broken Rotor Bars Based on EMD and Envelope Analysis. *Journal of China University Mining & Technology*, Vol.17, No. 2, pp.205-209.

Zhang, Pinjia; Du, Yi; Habetler, T.G.; Lu, Bin (2011). A Survey of Condition Monitoring and Protection Methods for Medium-Voltage Induction Motors. *IEEE Transactions on Industry Applications*, Vol.47, No.1, pp.34 - 46.

Pons-Llinares, J.; Antonino-Daviu, J.; Roger-Folch, J.; Morinigo-Sotelo, D.; Duque-Perez, O.; (2010). Eccentricity Diagnosis in Inverter Fed Induction Motors via the Analytic Wavelet Transform of Transient Currents. *XIX International Conference on Electrical Machines (ICEM)*, pp.1 -6.

10

Modeling, Simulation and Control of a Power Assist Robot for Manipulating Objects Based on Operator's Weight Perception

S. M. Mizanoor Rahman[1], Ryojun Ikeura[2] and Haoyong Yu[1]
1National University of Singapore
2Mie University
1Singapore
2Japan

1. Introduction

1.1 Power assist robot and its current applications

Power assist robot is a human-robot cooperation that extends human's abilities and skills in performing works (Kazerooni, 1993). Breakthrough in power assist robots was conceived in early 1960s with "Man-amplifier" and "Hardiman" (Kazerooni, 1993), but the progress of research on this significant field is not satisfactory yet. It is found through literature that power assist systems are currently being developed mainly for sick, physically disabled and old people as healthcare and rehabilitation supports (Kong *et al.*, 2009; Seki, Ishihara and Tadakuma, 2009). Few power assist systems have also been developed for other applications such as for lifting baby carriage (Kawashima, 2009), physical support for workers performing agricultural jobs (Tanaka *et al.*, 2008), hydraulic assist for automobiles (Liu *et al.*, 2009), skill-assist for manufacturing (Lee, Hara and Yamada, 2008), assisted slide doors for automobiles (Osamura *et al.*, 2008), assist-control for bicycle (Kosuge, Yabushita and Hirata, 2004), assist for sports training (Ding, Ueda and Ogasawara, 2008), etc.

1.2 Manipulating heavy objects in industries with power assist robots

We think that handling heavy objects, which is common and necessary in many industries, is another potential field of application of power assist robots. It is always necessary to move heavy objects in industries such as manufacturing and assembly, mining, construction, logistics and transport, disaster and rescue operations, forestry, agriculture etc. Manual manipulation of heavy objects is very cumbersome and it causes work-related disabilities and disorders such as back pain to humans. On the contrary, handling objects by autonomous systems may not provide required flexibility in many cases. Hence, it is thought that the uses of suitable human-robot cooperation systems such as power assist systems may be appropriate for handling heavy objects in industries. However, suitable power assist systems are not found in industries for this purpose because their design has not got much attention yet.

1.3 Weight illusion for power assist robots

A power assist robot reduces the perceived heaviness of an object manipulated with it (Kazerooni, 1993), as illustrated in Fig.1. Hence, load force (manipulative force tangential to grip surfaces) required to manipulate an object with a power assist robot should be lower than that required to manipulate the object manually. But, the limitations with the conventional power assist systems are that the operator cannot perceive the heaviness of the object correctly before manipulating it with the system and eventually applies excessive load force. The excessive load force results in sudden increase in acceleration, fearfulness of the operator, lack of maneuverability and stability, fatal accident etc. Fig.2 further explains the interaction processes and phenomena between a power assist robot and its operator for object manipulation. A few power assist systems are available for carrying objects (Doi *et al.*, 2007; Hara, 2007; Lee *et al.*, 2000; Miyoshi and Terashima, 2004). But, their safety, maneuverability, operability, naturalness, stability and other interactions with users are not so satisfactory because their controls do not consider human characteristics especially weight illusion and load force features.

Fig. 1. A human manipulates (lifts) an object with a power assist robot and feels a scaled-down portion of the weight.

1.4 Distinctions between unimanual and bimanual manipulation

It is noticed in practices in industries that workers need to employ one or two hands to manipulate objects and they decide this on the basis of object's physical features such as shape, size, mass etc. as well as of task requirements (Bracewell *et al.*, 2003; Giachritsis and Wing, 2008; Lum, Reinkensmeyer and Lehman, 1993;Rahman *et al.*, 2009a). We assume that weight perception, load force and object motions for unimanual manipulation may be different from that for bimanual manipulation, and these differences may affect modeling the control. Hence, it seems to be necessary to study unimanual weight perception, load force and motion features and to compare these to that for bimanual manipulation, and to reflect the differences in modeling the power-assist control. We studied distinctions between unimanual and bimanual manipulation in our previous works though it is still necessary to deeply look into their differences to make the control more appropriate (Rahman *et al.*, 2009a , 2011a).

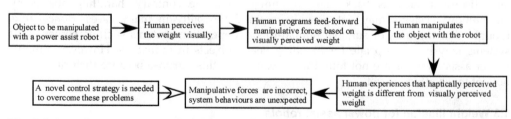

Fig. 2. Interaction processes and phenomena between robot and human when manipulating an object with a power assist robot.

1.5 Lifting, lowering and horizontal manipulation

In industries, workers need to transfer objects in different directions such as vertical lifting (lift objects from lower to higher position), vertical lowering (lower objects from higher to lower position), horizontal manipulation etc. in order to satisfy task requirements. We assume that maneuverability, heaviness perception, load force and motions for manipulating objects among these directions may be different from each other and these differences may affect the control and the system performances. Hence, it seems to be necessary to study object manipulation in all of these directions, compare them to each other, and to reflect the differences in the control (Rahman et al., 2010a, 2011a). However, such study has also not been carried out yet in detailed.We studied lifting objects in vertical direction in our previous works (Rahman *et al.*, 2009a, 2010c, 2011a), but manipulating objects in horizontal direction is still unaddressed though horizontal manipulation of objects is very common in practical fields. A few power-assist robotic systems consider manipulating objects in horizontal direction. But, they are not targetted to industrial applications and they have limitations in performances as they do not consider human characteristics in their control modeling (Hara, 2007).

1.6 The chapter summary

This chapter presents a power assist robot system developed for manipulating objects in horizontal direction in cooperation with human. Weight perception was included in robot dynamics and control. The robot was simulated for manipulating objects in horizontal direction. Optimum maneuverability conditions for horizontal manipulation of objects were determined and were compared to that for vertical lifting of objects. Psychophysical relationships between actual and perceived weights were determined, and load forces and motion features were analyzed for horizontal manipulation of objects. Then, a novel control scheme was implemented to reduce the excessive load forces and accelerations, and thus to improve the system performances. The novel control reduced the excessive load forces and accelerations for horizontal manipulation of objects, and thus improved the system performances in terms of maneuverability,safety, operability etc. We compared our results to that of related works. Finally, we proposed to use the findings to develop human-friendly power assist robots for manipulating heavy objects in various industries.

This chapter provides information to the readers about the power assist robot system- its innovative mechanical design, dynamics, modeling, control, simulation, application etc. Thus this chapter introduces a new area of applications of power assist robot systems and also introduces innovations in its dynamics, modeling, control etc. On the other hand, the readers will get a detailed explanation and practical example of how to use Matlab/Simulink to develop and simulate a dynamic system (e.g., a power assist robot system). The readers will also receive a practical example of how to measure and evaluate human factors subjectively for a technical domain (e.g., a power assist robot system). As a whole, this chapter will enrich the readers with novel concepts in robotics and control technology, Matlab/Simulink application, human factors/ergonomics, psychology and psychophysics, biomimetics, weight perception, human-robot/machine interaction, user interface design, haptics, cognitive science, biomechanics etc. The contents of this chapter were also compared to related works available in published literatures. Thus, the readers will get a collection of all possible works related and similar to the contents of this chapter.

2. The experimental robotic system: Configuration, dynamics and control

2.1 Configuration

We developed a 1-DOF (horizontal translational motion) power assist robot system using a ball screw assembly actuated by an AC servomotor (Type: SGML-01BF12, made by Yaskawa, Japan). The ball screw assembly and the servomotor were coaxially fixed on a metal board and the board was horizontally placed on a table. Three rectangular boxes were made by bending aluminum sheets (thickness: 0.5 mm).These boxes were horizontally manipulated with the power assist robot system and were called the power-assisted objects (PAOs). A PAO (box), at a time, could be tied to the ball nut (linear slider) of the ball screw assembly through a force sensor (foil strain gauge type, NEC Ltd.) and be manipulated by a human subject. The dimensions (length x width x height) of the boxes were 16 x 6 x 5cm, 12 x 6 x 5cm and 8.6 x 6 x 5cm for the large, medium and small size respectively. The bottom, left and right sides of each PAO were open. The complete experimental setup of the power assist robot system is depicted in Fig.3. The servodrive receives a command signal (voltage signal) from the controller through a D/A converter, amplifies the signal, and transmits electric current to the servomotor in order to produce motion proportional to the commanded signal. The position sensor (encoder with counter) reports (pulse signal) object's actual displacement back to the servodrive. The servodrive then compares the actual displacement to the desired displacement. It then alters the commanded signal to the motor so as to correct for any error in the displacement. Human force is sensed by the force sensor attached between the ball nut and the object. The force is sensed as voltage signal, amplified by the amplifier and then sent to the control system via an A/D converter. Human force gives only motion (acceleration) to the object.

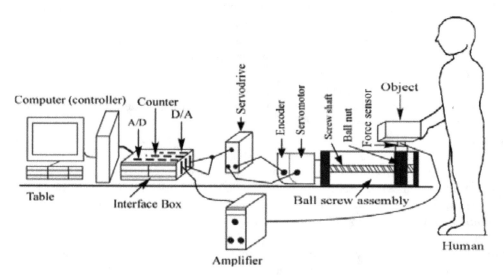

Fig. 3. Experimental setup of the 1-DOF power assist robot system.

2.2 Dynamics

According to Fig.4, the dynamics of the PAO when it is manipulated horizontally by a subject with the power assist robot system is described by Eq.(1), where F_o=mg. If we include our hypothesis in the dynamics, then Eq.(1) changes to Eq.(2). Both m_1 and m_2 stand for mass, but m_1 forms inertial force and m_2 forms gravitational force, $m_1 \neq m_2 \neq m, m_1 \ll m, m_2 \ll m, |m_1\ddot{x}_d| \neq |m_2 g|$. A difference between m_1 and m_2 arises due to the difference between human's perception and reality regarding the heaviness of the object manipulated with the power assist system (Kazerooni, 1993).

$$m\ddot{x}_d = f_h + F_0 \tag{1}$$

$$m_1\ddot{x}_d = f_h + m_2 g. \tag{2}$$

Where,
m = Actual mass of the object
x_d = Desired displacement of the object f_h = Load force applied by the subject
g = Acceleration of gravity

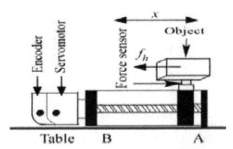

Fig. 4. Dynamics of 1-DOF power assist system for horizontal manipulation of objects. The PAO tied to the force sensor is moved by the subject from 'A' to 'B' position.

2.3 Control
We derived Eqs. (3)- (5) from Eq. (2). We then diagrammed the power-assist control based on Eqs.(3)-(5), which is shown in Fig.5. Eq. (3) gives the desired acceleration. Then, Eq. (3) is integrated and the integration gives the desired velocity $(\dot{x}_d)$. Then, the velocity is integrated and the integration gives the desired displacement (x_d).

$$\ddot{x}_d = \frac{1}{m_1}(f_h + m_2 g) \tag{3}$$

$$\dot{x}_d = \int \ddot{x}_d \, dt \tag{4}$$

$$x_d = \int \dot{x}_d \, dt \tag{5}$$

$$\dot{x}_c = \dot{x}_d + G(x_d - x) \tag{6}$$

If the system is simulated using Matlab/Simulink in the velocity control mode of the servomotor, the commanded velocity $(\dot{x}_c)$ to the servomotor is calculated by Eq. (6). The commanded velocity is provided to the servomotor through the D/A converter. During simulation, the servodrive determines the error displacement signal by comparing the actual displacement to the desired displacement and generates the control.
The following three types of control methods are usually used in power assist systems:-
1. Position based impedance control
2. Torque/force based impedance control
3. Force control

Position based impedance control and torque/force based impedance control produce good results. Results may be different for force control for reducing excessive force. Our control as introduced above is limited to position based impedance control. We used position based impedance control for the following reasons (advantages):

1. Position based impedance control automatically compensates the effects of friction, inertia, viscosity etc. In contrast, these effects are needed to consider for force control, however, it is very difficult to model and calculate the friction force for the force control. Dynamic effects, nonlinear forces etc. affect system performances for force control for multi-degree of freedom system.
2. Ball-screw gear ratio is high and actuator force is less for position control. However, the opposite is true for the force control.
3. It is easy to realize the real system for the position control for high gear ratio. However, the opposite is true for the force control.

However, there are some disadvantages of position control as the following:

1. Instability is high in position control. If we see Fig.5 we find that a feedback loop is created between x and f_h when human touches/grasps the object for manipulation. This feedback effect causes instability. In contrast, force control has less or no stability problem.
2. Motor system delay affects the stability more intensively for position control.
3. Value of G and velocity control also can compensate the effects of friction, inertia, viscosity etc., but the effects are not compensated completely, which may affect human's weight perception.

If the difference between m and m_1 is very large i.e., if (m-m_1) is very big, the position control imposes very high load to the servomotor that results in instability, which is not so intensive for force control. Force control is better in some areas, but position control is better in some other areas. However, position control is to be effective for this chapter. Force control may be considered in near future.

The control shown in Fig.5 is not so complicated. However, there is novelty in this control that human's perception is included in this control. Again, another novel control strategy is also derived from this control that includes human features. This control can be recognized as an exemplary and novel control for human interactive robot control.

3. Experiment 1: Analyzing maneuverability,heaviness perception, force and motion features

3.1 Subjects

Ten mechanical engineering male students aged between 23 and 30 years were nominated to voluntarily participate in the experiment. The subjects were believed to be physically and mentally healthy. The subjects did not have any prior knowledge of the hypothesis being tested. Instructions regarding the experiment were given to them, but no formal training was arranged.

3.2 Evaluation criteria for power assist robot systems

Power assist can be defined as augmenting the ability or adjusting to the situation when human operates and works. In particular, in case of supporting for elderly and disabled people, the purpose of power assist is improvement of QOL (Quality of Life), that is, support for daily life. It has two meanings. One is support for self-help and the other is support for caring. The former is to support self-sustained daily life, and the latter is to decrease burden of caregiver.

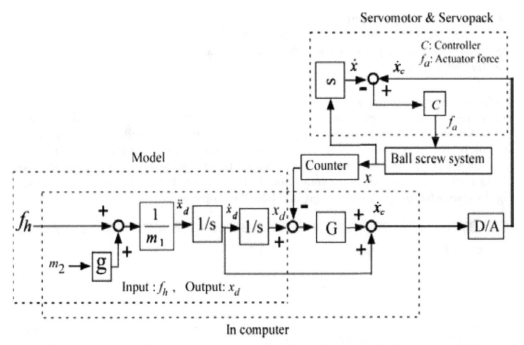

Fig. 5. Block diagram of the power-assist control, where G denotes feedback gain, D/A indicates D/A converter and x denotes actual displacement. Feedback position control is used with the servomotor in velocity control mode.

There are two requirements for power assist systems. The first requirement is amplification of human force, assistance of human motion etc. This is the realization of power assist itself, and there may have problems such as its realization method and stability of human-robot system. The second is safety, sense of security, operability, ease of use etc. This requirement does not appear as specific issue comparing to the first requirement, and it is difficult to be taken into account. However, in order to make power assist systems useful, the second requirement is more important than the first requirement. In case of lifting objects, we think that the main requirements for the power assist systems are maneuverability, safety and stability. Again, these requirements are interrelated where maneuverability plays the pivotal role.

Some basic requirements of a power assist system regarding its maneuverability have been mentioned by Seki, Iso and Hori (2002).However, we thought that only the light (less force required), natural and safe system can provide consistent feelings of ease of use and comfort though too light system may be unsafe, uneasy and uncomfortable. Hence, we considered operator's ease of use and comfort as the evaluation criteria for maneuverability of the power assist robot system.

3.3 Objectives

Objectives of experiment 1 were to (i) determine conditions for optimum maneuverability, (ii) determine psychophysical relationships between actual and perceived weights, (iii) analyze load force and determine excess in load force, (iv) analyze object's motions-displacement, velocity and acceleration etc. for manipulating objects with the power assist robot system in horizontal direction.

3.4 Design of the experiment

Independent variables were m_1 and m_2 values, and visual object sizes. Dependent variables were maneuverability, perceived weight, load force, and object's motions (displacement, velocity and acceleration).

3.5 Experiment procedures

The system shown in Fig.5 was simulated using Matlab/Simulink (solver: ode4, Runge-Kutta; type: fixed-step; fundamental sample time: 0.001s) for twelve m_1 and m_2 sets (Table 1) separately. The ranges of values of m_1 and m_2 were nominated based on our experience. The program for simulation is shown in Fig.6. We set the parameters of three custom-derived blocks such as counter, D/A converter and A/D converter before the simulation started. Fig. 7 shows what the parameters were and how they were set.

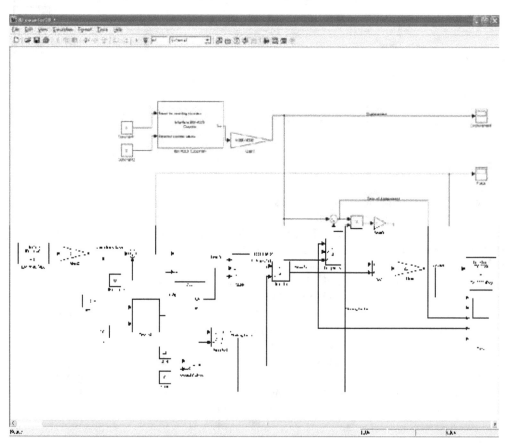

Fig. 6. The program for the simulation

The subject manipulated (from 'A' to 'B' as in Fig.4, distance between 'A' and 'B' was about 0.12 m) each size object with the robot system once for each m_1 and m_2 set separately. The task required the subject to manipulate the object approximately 0.1m, maintain the object for 1-2 seconds and then release the object. For each trial (for each m_1 and m_2 set for each size object), the subject subjectively evaluated the system for maneuverability as any one of the following alternatives:-

1.   Very Easy & Comfortable (score: +2)
2.   Easy & Comfortable (score: +1)
3.   Borderline (score: 0)
4.   Uneasy & Uncomfortable (score: -1)
5.   Very Uneasy & Uncomfortable (score: -2)

All subjects evaluated the system for maneuverability as above for small, medium, large object independently for each m_1 and m_2 set. Load force and motions data were recorded separately for each trial.

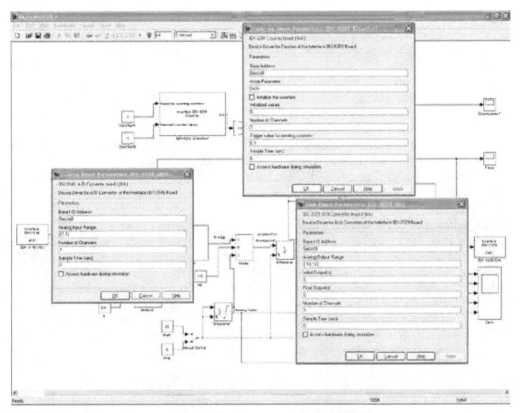

Fig. 7. Setting appropriate parameters for the custom-derived blocks.

Each subject after each trial also manually manipulated a reference-weight object horizontally on a table using right hand alone for reference weights. Weight of the reference-weight object was sequentially changed in a descending order starting from 0.1 kg and ending at 0.01 kg maintaining an equal difference of 0.01 kg i.e., 0.1, 0.09,...0.02, 0.01kg.The subject thus compared the perceived weight of the PAO to that of the reference-weight object and estimated the magnitude of the perceived weight following the psychophysical method 'constant stimuli'. Appearance of PAO and reference-weight object were the same.

m_1 (kg)	2.0	1.5	1.0	0.5
m_2 (kg)	0.09	0.06	0.03	

Table 1. Values of variables for the simulation

3.6 Experiment results
3.6.1 Optimum maneuverability

Mean evaluation scores of the system regarding its maneuverability for 12 m_1 and m_2 sets for each size object were determined separately. Table 2 shows the mean evaluation scores for the medium size object. Similar scores were also determined for large and small size objects. The results reveal that maneuverability is not affected by visual size of object. The reason may be that human evaluates maneuverability using haptic senses where visual size cue has no influence. However, haptic cues might influence the maneuverability.

The table shows that ten m_1 and m_2 sets got positive scores whereas the remaining two sets got negative scores. Results show that m_1=0.5kg, m_2=0.03kg and m_1=1kg, m_2=0.03kg got the highest scores. Hence, optimum maneuverability may be achieved at any one of these two conditions. We think that a unique and single condition for optimum maneuverability could be determined if more values of m_1 and m_2 were used for the simulation. The subjects felt very easy and comfortable to manipulate objects with the power assist system only when m_1=0.5kg, m_2=0.03kg and m_1=1kg, m_2=0.03kg. This is why these two sets were declared as the optimum conditions for maneuverability. Here, optimality was decided based on human's feelings following heuristics.

These findings indicate the significance of our hypothesis that we would not be able to sort out the positive sets (satisfactory level of maneuverability) of values of m_1 and m_2 from the negative sets (unsatisfactory level of maneuverability) of values of m_1 and m_2 for different sizes of objects unless we thought $m_1 \neq m_2 \neq m, m_1 \ll m, m_2 \ll m, m_1\ddot{x}_d \neq m_2 g$.

We see that the optimum/best sets are also the sets of the smallest values of m_1 and m_2 in this experiment. If much smaller values of m_1 and m_2 are chosen randomly, the perceived heaviness may further reduce, but it needs to clarify whether or not this is suitable for human psychology. Again, in zero-gravity or weightless condition when m_2=0, the object is supposed to be too light as it was studied by Marc and Martin (2002) in actual environment and by Dominjon et al. (2005) in virtual environment. It was found that the zero-gravity is not feasible because the human loses some haptic information at zero-gravity that hampers human's weight perception ability (Rahman et al., 2009b).It is still not known whether the optimum sets are optimum only for the particular conditions of this experiment or they will persist as the optimum for all conditions in practical uses in industries.

m_1	m_2	Mean maneuverability score
1	0.06	+0.83(0.41)
2	0.06	+0.33(1.21)
0.5	0.03	+2.0 (0)
1	0.03	+2.0 (0)
1.5	0.03	+1.5 (0.55)
2	0.09	-0.17(0.98)
0.5	0.06	+1.0 (0)
1.5	0.09	-0.17(0.98)
0.5	0.09	+0.17(0.75)
1	0.09	+1.0 (0.63)
1.5	0.06	+0.67(0.52)
2	0.03	+1.17(0.41)

Table 2. Mean maneuverability scores with standard deviations (in parentheses) for the medium size object

3.6.2 Relationship between actual and perceived weight

We determined the mean perceived weight for each size object separately for m_1=0.5kg, m_2=0.03kg (condition 1) and m_1=1kg, m_2=0.03 kg (condition 2) as shown in Fig.8. We assumed m_2 as the actual weight of the power-assisted object. It means that the actual weight was 0.03kg or 0.2943 N for each size object for the two m_1 and m_2 sets. We compared the perceived weights of Fig.8 to the actual weight (0.2943 N) for each size object for m_1=0.5kg, m_2=0.03kg and m_1=1kg, m_2=0.03 kg. The figure shows and we also found in our previous research that m_1 does not affect weight perception, but m_2 does affect (Rahman et al., 2009a, 2011a). We also see that visual object sizes do not affect weight perception (Gordon et al., 1991). Results for two-way (visual object size, subject) analyses of variances separately analyzed on perceived weights for the two m_1 and m_2 sets showed that variations due to object sizes were insignificant ($F_{2, 18}$ <1 for each m_1 and m_2 set).The reason may be that subjects estimated perceived weights using haptic cues where visual cues had no influences. Variations among subjects were also found statistically insignificant ($F_{9,18}$<1 for each m_1 and m_2 set).

The actual weight of the object was 0.2943 N, but the subjects felt about 0.052 N when the object was manipulated with the power assist robot system in horizontal direction. Hence, the results reveal that the perceived weight is about 18% of the actual weight if an object is manipulated horizontally with a power assist robot system. Its physical meaning is that the perceived weight of an object manipulated with power-assist in horizontal direction is 18% of the perceived weight of the same object manipulated in horizontal direction manually. This happens because the power assist robot system reduces the perceived weight through its assistance to the user. It is a well-known concept that a power assist robot system reduces the feeling of weight. However, it was not quantified. This research quantified the weight attenuation for horizontal manipulation of objects with the power assist robot system. As we found in our previous research, the perceived weight reduces to 40% and 20% of the actual weight if the object is vertically lifted (Rahman et al., 2011a) or vertically lowered (Rahman et al., 2011b) respectively. The weight perception is less for horizontal manipulation as the gravity force is compensated.

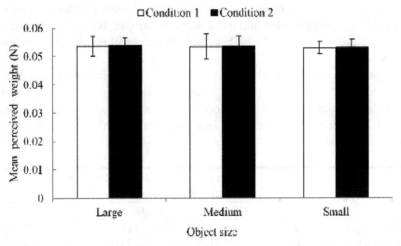

Fig. 8. Mean (n=10) perceived weights for different object sizes for condition 1 (m_1=0.5kg, m_2=0.03kg) and condition 2 (m_1=1kg, m_2=0.03 kg).

3.6.3 Force analysis

The time trajectory of load force for a typical trial is shown in Fig.9. We derived the magnitude of peak load force (PLF) for each object size   for condition 1 (m_1=0.5kg, m_2=0.03kg) and condition 2 (m_1=1kg, m_2=0.03 kg) separately and determined the mean PLFs. The results are shown in Table 3. Results show that mean PLFs for condition 2 are slightly larger than that for condition 1. We found previously that both m_1 and m_2 are linearly proportional to peak load force. However, m_1 affects load force, but it does not affect weight perception. On the other hand, m_2 affects both load force and weight perception (Rahman *et al.*, 2011a). Here, we assume that larger m_1 in condition 2 has produced larger load force.

We have already found that subjects feel the best maneuverability at m_1=0.5kg, m_2=0.03kg and m_1=1kg, m_2=0.03 kg. On the other hand, actually required PLF to manipulate the power-assisted object should be slightly larger than the perceived weight (Gordon *et al.*, 1991), which is 0.052 N. We compared the perceived weights from Fig.8 to the PLFs (Table 3) for the large, medium and small objects and determined the excess in PLFs. The results show that subjects apply load forces that are extremely larger than the actually required load forces for condition 1 and 2. We also see that the magnitudes of PLFs are proportional to object sizes (Gordon *et al.*, 1991). We assume that the excessive load forces create problems in terms of maneuverability, safety, motions etc. that we discussed in the introduction.

3.6.4 Motion analysis

Fig.9 shows trajectories of displacement,velocity and acceleration for a typical trial. The figure shows that the time trajectories of load force and object's acceleration are synchronized i.e., when load force reaches the peak; acceleration also reaches the peak and so on. However, the trajectory of displacement is different from that of load force and acceleration i.e., the displacement is not entirely synchronized with load force and acceleration. Hence, we see that there is a time delay between PLF (peak acceleration as well) and peak displacement. Previously we assumed that the time delay is caused due to a delay in position sensing (Rahman et al., 2010b), but this research reveals that the time delay may be caused by the combined effects of the time constant of the position sensor and the delay in adjusting the situation and motions by the subject. We also assume that the time delay may cause the feeling of reduced heaviness of the object manipulated with the power assist robot system.

We derived peak velocity and peak acceleration for each trial and determined their means for each object size in each condition separately as shown in Table 4 and Table 5 respectively. The results show that the velocity and accelerations are large. We assume that the large peak load forces have resulted in large accelerations that are harmful to the system in terms of maneuverability, safety, motions etc.

m_1, m_2 sets	Mean PLFs (N) with standard deviations (in parentheses) for different object sizes		
	Large	Medium	Small
m_1=0.5kg, m_2=0.03kg	2.9131(0.1307)	2.6020(0.1151)	2.4113(0.1091)
m_1=1.0kg, m_2=0.03kg	2.9764(0.2009)	2.6554(0.1552)	2.4602(0.1367)

Table 3. Mean peak load forces for different conditions for different object sizes

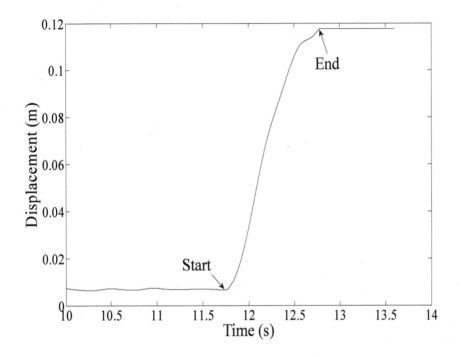

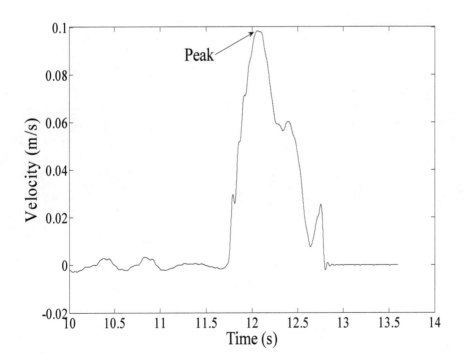

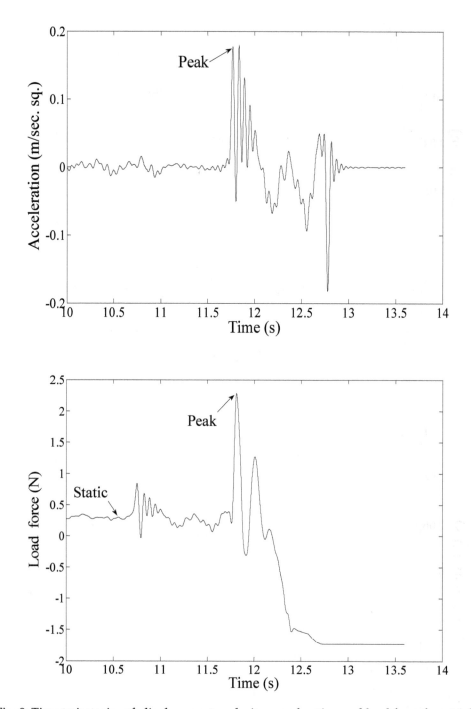

Fig. 9. Time trajectories of displacement , velocity , acceleration and load force for a trial when a subject manipulated the small size PAO with the system at condition 1 (m_1=0.5kg, m_2=0.03kg).

Object size	Mean peak velocity (m/s)	
	m_1=0.5kg, m_2=0.03kg	m_1=1.0kg, m_2=0.03kg
Large	0.1497(0.0149)	0.1557(0.0209)
Medium	0.1345(0.0157)	0.1399(0.0122)
Small	0.1098(0.0121)	0.1176(0.0119)

Table 4. Mean peak velocity with standard deviations (in parentheses) for different object sizes for different conditions

Object size	Mean peak acceleration (m/s²)	
	m_1=0.5kg, m_2=0.03kg	m_1=1.0kg, m_2=0.03kg
Large	0.2309 (0.0901)	0.2701 (0.0498)
Medium	0.2282 (0.0721)	0.2542(0.0153)
Small	0.1887(0.0298)	0.2134(0.0525)

Table 5. Mean peak accelerations with standard deviations (in parentheses) for different object sizes for different conditions.

4. Experiment 2: Improving system performances by a novel control

4.1 Experiment

Table 3 and Table 5 show that subjects apply too excessive load forces and accelerations that cause problems as we discussed in section 1. Experiment 2 attempted to reduce excessive load forces and accelerations by applying a novel control method.

The novel control was such that the value of m_1 exponentially declined from a large value to 0.5kg when the subject manipulated the PAO with the system and the command velocity of Eq.(6) exceeded a threshold. We found previously that load force is linearly proportional to m_1 and we also found that subjects do not feel the change of m_1 (Rahman et al., 2011a). Hence, reduction in m_1 would also reduce the load force proportionally. Reduction in load force would not adversely affect the relationships of Eq. (2) because the subjects would not feel the change of m_1. It means that Eq. (7) and Eq. (8) were used for m_1 and m_2 respectively to modify the control of Fig.5. The digit 6 in Eq. (7) was determined by trial and error. The novel control is illustrated in Fig.10 as a flowchart. The procedures for experiment 2 were the same as that for the experiment 1, but m_1 and m_2 were set as m_1=6*e^{-6t} + 0.5, m_2=0.03 (condition 1.a) and m_1=6*e^{-6t} + 1.0, m_2=0.03 (condition 2.a) for the simulation. Program for the simulation is shown in Fig.11. We here ignore presenting the simulation details for m_1=6*e^{-6t} + 1.0, m_2=0.03 because the concept and procedures for m_1=6*e^{-6t} + 0.5, m_2=0.03 and m_1=6*e^{-6t} + 1.0, m_2=0.03 are the same.

$$m_1 = 6 * e^{-6t} + 0.5 \tag{7}$$

$$m_2 = 0.03 \tag{8}$$

The system performances were broadly expressed through several criteria such as motion, object mobility, naturalness, stability, safety, ease of use etc., and in each trial in each scheme, the subjects subjectively evaluated (scored) the system using a 7-point bipolar and equal-interval scale as follows:
1. Best (score: +3)
2. Better (score: +2)

3. Good (score: +1)
4. Alike (score: 0)
5. Bad (score:-1)
6. Worse (score:-2)
7. Worst (score:-3)

4.2 Experiment results
4.2.1 Reduction in peak load forces and peak accelerations
We compared the mean PLFs of experiment 2 conducted at $m_1=6 * e^{-6t} + 0.5$, $m_2=0.03$ and $m_1=6 * e^{-6t} + 1.0$, $m_2=0.03$ to that of experiment 1 conducted at $m_1=0.5$, $m_2=0.03$ and $m_1=1.0$, $m_2=0.03$. The findings are shown in Table 6. Findings show that PLFs reduced significantly due to the control modification.

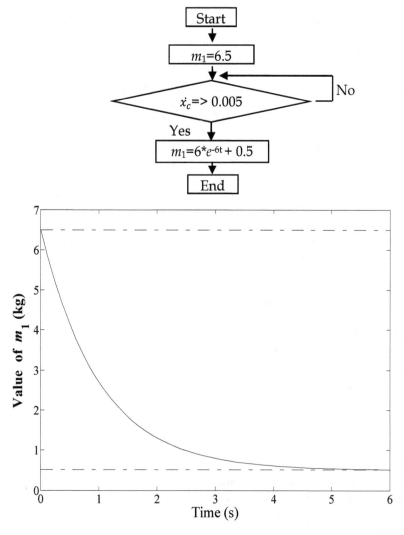

Fig. 10. Flowchart and hypothetical trajectory of inertial mass for the novel control technique.

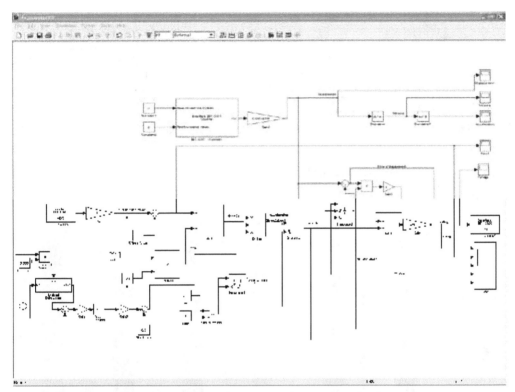

Fig. 11. The program for the simulation for the novel control method

Mean peak accelerations for different object sizes after the control modification are shown in Table 7. The results show, if we compare these to that of Table 5, that the peak accelerations reduced due to control modification. The reason may be that the reduced peak load forces after the control modification reduced the accelerations accordingly. The velocity did not change significantly due to the control modification.

4.2.2 Improvement in system performances
We determined the mean evaluation scores for the three objects separately. Fig.12 shows the mean scores for the small size object for both conditions. The scores for the large and medium size objects in each condition were almost the same as that shown in the figure. It means that the novel control was effective in improving the system performances.

m_1, m_2 sets	Mean PLFs (N) with standard deviations (in parentheses) for different object sizes		
	Large	Medium	Small
$m_1=6 * e^{-6t} + 0.5$, $m_2=0.03$	1.3569 (0.1154)	1.1123(0.0821)	0.9901(0.0910)
$m_1=6 * e^{-6t} + 1.0$, $m_2=0.03$	1.8646 (0.1707)	1.5761(0.1071)	1.0990 (0.0885)

Table 6. Mean peak load forces for different conditions for different object sizes after the control modification

Object size	Mean peak acceleration (m/s²)	
	$m_1=6 * e^{-6t} + 0.5$, $m_2=0.03$	$m_1=6 * e^{-6t} + 1.0$, $m_2=0.03$
Large	0.1234 (0.0403)	0.1404 (0.0302)
Medium	0.1038 (0.0233)	0.1220 (0.0107)
Small	0.0884 (0.0311)	0.1008 (0.0164)

Table 7. Mean peak accelerations with standard deviations (in parentheses) for different object sizes for different conditions after the control modification

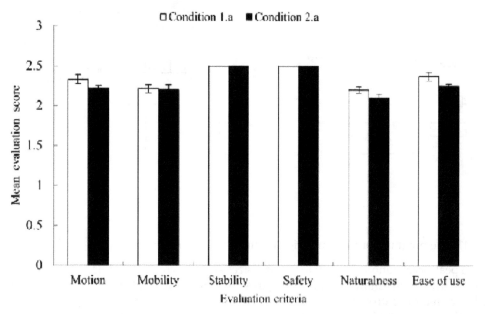

Fig. 12. Mean performance evaluation scores for small size object for condition 1.a ($m_1=6 *$ e^{-6t} + 0.5, $m_2=0.03$) and condition 2.a ($m_1=6 *$ e^{-6t} + 1.0, $m_2=0.03$) after the control modification.

5. Conclusions

In this chapter, we presented a 1-DOF power assist robot system for manipulating objects by human subjects in horizontal direction. We included human features in the robot dynamics and control. We determined optimum maneuverability conditions for manipulating objects with the robot system. We also determined psychophysical relationships between actual and perceived weights for manipulating objects with the robot system. We analyzed weight perception, load forces and motion characteristics. We implemented a novel control method based on weight perception, load forces and motion characteristics that improved the system performances through reducing the peak load forces and peak accelerations. The findings may help develop human-friendly power assist robot devices for manipulating heavy objects in industries such as manufacturing and assembly, mining, logistics and transport, construction etc. This chapter also provides a vivid example to the readers of how Matlab/Simulink is used to model and develop control system and interfaces between hardware and software for simulation and control of a robotic system. The findings of this chapter are novel and they enhance the state-of-the-art knowledge and applications of robotics, control system, simulation, Matlab/Simulink, psychology, human factors etc. We will verify the results using heavy objects and real robotic systems in near future. The system will be upgraded to multi-degree of freedom system. Distinctions in weight perception, load forces and motion characteristics between unimanual and bimanual manipulation of objects in horizontal direction will be investigated.

6. Acknowledgment

The authors are thankful to the Ministry of Education, Culture, Sports, Science and Technology of Japan for financial supports.

7. References

Bracewell, R.M., Wing, A.M., Scoper, H.M., Clark, K.G. (2003) 'Predictive and reactive co-ordination of grip and load forces in bimanual lifting in man', *European Journal of Neuroscience*, Vol.18, No.8, pp.2396-2402.

Ding, M., Ueda, J., Ogasawara, T. (2008) 'Pinpointed muscle force control using a power-assisting device: system configuration and experiment', *In Proc. of IEEE Int. Conf. on Biomedical Robotics and Biomechatronics*, pp.181 – 186.

Doi, T., Yamada, H., Ikemoto, T. and Naratani, H. (2007) 'Simulation of pneumatic hand crane type power assist system', *In Proc. of SICE Annual Conf.*, pp. 2321 – 2326.

Dominjon, L., Lécuyer, A., Burkhardt, J.M., Richard, P. and Richir, S. (2005) 'Influence of control/display ratio on the perception of mass of manipulated objects in virtual environments', *In Proc. of IEEE Virtual Reality*, pp.19-25.

Giachritsis, C. and Wing, A. (2008) 'Unimanual and bimanual weight discrimination in a desktop setup', M. Ferre (Ed.): *EuroHaptics 2008*, LNCS 5024, pp. 378–382.

Gordon, A.M., Forssberg, H., Johansson, R.S., Westling, G. (1991) 'Visual size cues in the programming of manipulative forces during precision grip', *Exp. Brain Research*, Vol. 83, No.3, pp. 477–482.

Hara, S. (2007) 'A smooth switching from power-assist control to automatic transfer control and its application to a transfer machine', *IEEE Trans. on Industrial Electronics*, Vol. 54, No. 1, pp.638-650.

Kawashima, T. (2009) 'Study on intelligent baby carriage with power assist system and comfortable basket', *J. of Mechanical Science and Technology*, Vol.23, pp.974-979.

Kazerooni, H. (1993) 'Extender: a case study for human-robot interaction via transfer of power and information signals', *In Proc. of IEEE Int. Workshop on Robot and Human Communication*, pp.10-20.

Kong, K., Moon, H., Hwang, B., Jeon, D., Tomizuka, M. (2009) 'Impedance compensation of SUBAR for back-drivable force-mode actuation', *IEEE Trans. on Robotics*, Vol. 25, Issue: 3, pp.512 – 521.

Kosuge,K., Yabushita, H., Hirata, Y. (2004) 'Load-free control of power-assisted cycle', *In Proc. of IEEE Technical Exhibition Based Conference on Robotics and Automation*, pp.111-112.

Lee, H., Takubo, T., Arai, H. and Tanie, K. (2000) 'Control of mobile manipulators for power assist systems', *Journal of Robotic Systems*, Vol.17, No.9, pp.469-477.

Lee, S., Hara, S., Yamada, Y. (2008) 'Safety-preservation oriented reaching monitoring for smooth control mode switching of skill-assist', *In Proc. of IEEE Int. Conf. on Systems, Man and Cybernetics*, pp.780–785.

Liu, G., Yan, Y., Chen, J., Na, T. (2009) 'Simulation and experimental validation study on the drive performance of a new hydraulic power assist system', *In Proc. of IEEE Intelligent Vehicles Symposium*, pp.966-970.

Lum, P.S., Reinkensmeyer, D.J., Lehman, S.L. (1993) 'Robotic assist devices for bimanual physical therapy: preliminary experiments', *IEEE Trans. on Rehabilitation Engineering*, Vol.1, No.3, pp.185-191.

Marc, O.E., and Martin, S.B. (2002) 'Humans integrate visual and haptic information in a statistically optimal fashion', *Nature*, Vol. 415, No.6870, pp.429-433.

Miyoshi, T. and Terashima, K. (2004) 'Development of vertical power-assisted crane system to reduce the operators' burden', *In Proc. of IEEE Int. Conference on Systems, Man and Cybernetics*, Vol.5, pp. 4420 - 4425.

Osamura, K., Kobayashi, S., Hirata, M., Okamoto, H. (2008) 'Power assist control for slide doors using an ideal door model', *In Proc. of IEEE Int. Symposium on Industrial Electronics*, pp. 1293 - 1299.

Rahman,S.M.M., Ikeura, R., Nobe, M., Sawai,H. (2009a) 'Design and control of a 1DOF power assist robot for lifting objects based on human operator's unimanual and bimanual weight discrimination', *In Proc. of IEEE Int. Conf. on Mechatronics and Automation*,pp.3637-3644.

Rahman, S.M.M., Ikeura, R., Nobe, M., Sawai, H. (2009b) 'A psychophysical model of the power assist system for lifting objects', *In Proc. of IEEE Int. Conf. on Systems, Man, and Cybernetics*, pp.4125-4130.

Rahman, S.M.M., Ikeura, R., Nobe, M., Sawai, H. (2010a) 'Study on optimum maneuverability in horizontal manipulation of objects with power-assist based on weight perception', *In Proc. of SPIE*, Vol. 7500, 75000P.

Rahman, S.M.M., Ikeura, R., Nobe, M., Sawai, H. (2010b) 'Displacement-load force-perceived weight relationships in lifting objects with power-assist', *In Proc. of SPIE*, Vol. 7500, 75000S.

Rahman, S.M.M., Ikeura, R., Nobe, M., Sawai, H. (2010c), 'Controlling a power assist robot for lifting objects considering human's unimanual, bimanual and cooperative weight perception', *In Proc. of IEEE Int. Conf. on Robotics and Automation*, pp.2356-2362.

Rahman, S.M.M., Ikeura, R., Hayakawa, S. and Sawai, H. (2011a) 'Design guidelines for power assist robots for lifting heavy objects considering weight perception, grasp differences and worst-cases', *Int. J. Mechatronics and Automation*, Vol. 1, No. 1, pp.46–59.

Rahman, S.M.M., Ikeura, R., Ishibashi, S., Hayakawa, S., Sawai, H., Yu, H. (2011b) 'Lowering objects manually and with power-assist: distinctions in perceived heaviness, load forces and object motions', *In Proc. of 4th IEEE International Conference on Human System Interactions, 19-21 May, 2011, Yokohama, Japan, pp.129-135*.

Seki,H.,Iso,M.,Hori,Y.(2002) 'How to design force sensorless power assist robot considering environmental characteristics-position control based or force control based-', *In Proc. of Annual Conf. of IEEE Industrial Electronics Society*,Vol.3,pp.2255-2260.

Seki,H.,Ishihara,K.,Tadakuma,S.(2009) 'Novel regenerative braking control of electric power-assisted wheelchair for safety downhill road driving', *IEEE Trans. on Industrial Electronics*, Vol. 56, No. 5, pp. 1393-1400.

Tanaka, T., Satoh, Y., Kaneko, S., Suzuki, Y., Sakamoto, N. and Seki, S. (2008) 'Smart suit: soft power suit with semi-active assist mechanism – prototype for supporting waist and knee joint', In Proc. of Int. Conf. on Control, Automation and Systems, pp. 2002-2005.

Permissions

All chapters in this book were first published by InTech Open; hereby published with permission under the Creative Commons Attribution License or equivalent. Every chapter published in this book has been scrutinized by our experts. Their significance has been extensively debated. The topics covered herein carry significant findings which will fuel the growth of the discipline. They may even be implemented as practical applications or may be referred to as a beginning point for another development.

The contributors of this book come from diverse backgrounds, making this book a truly international effort. This book will bring forth new frontiers with its revolutionizing research information and detailed analysis of the nascent developments around the world.

We would like to thank all the contributing authors for lending their expertise to make the book truly unique. They have played a crucial role in the development of this book. Without their invaluable contributions this book wouldn't have been possible. They have made vital efforts to compile up to date information on the varied aspects of this subject to make this book a valuable addition to the collection of many professionals and students.

This book was conceptualized with the vision of imparting up-to-date information and advanced data in this field. To ensure the same, a matchless editorial board was set up. Every individual on the board went through rigorous rounds of assessment to prove their worth. After which they invested a large part of their time researching and compiling the most relevant data for our readers.

The editorial board has been involved in producing this book since its inception. They have spent rigorous hours researching and exploring the diverse topics which have resulted in the successful publishing of this book. They have passed on their knowledge of decades through this book. To expedite this challenging task, the publisher supported the team at every step. A small team of assistant editors was also appointed to further simplify the editing procedure and attain best results for the readers.

Apart from the editorial board, the designing team has also invested a significant amount of their time in understanding the subject and creating the most relevant covers. They scrutinized every image to scout for the most suitable representation of the subject and create an appropriate cover for the book.

The publishing team has been an ardent support to the editorial, designing and production team. Their endless efforts to recruit the best for this project, has resulted in the accomplishment of this book. They are a veteran in the field of academics and their pool of knowledge is as vast as their experience in printing. Their expertise and guidance has proved useful at every step. Their uncompromising quality standards have made this book an exceptional effort. Their encouragement from time to time has been an inspiration for everyone.

The publisher and the editorial board hope that this book will prove to be a valuable piece of knowledge for researchers, students, practitioners and scholars across the globe.

List of Contributors

Lubomír Brančík
Brno University of Technology, Czech Republic

Gabriel Hulkó, Cyril Belavý, Gergely Takács, Pavol Buček and Peter Zajíček
Institute of Automation, Measurement and Applied, Informatics Faculty of Mechanical Engineering, Center for Control of Distributed Parameter Systems, Slovak University of Technology Bratislava, Slovak Republic

A. K. Parvathy and R. Devanathan
Hindustan Institute of Technology and Science, Chennai, India

Mohamed Amine Fakhfakh, Moez Ayadi, Ibrahim Ben Salah and Rafik Neji
University of Sfax/Sfax, Tunisia

Conghui Liang, Marco Ceccarelli and Giuseppe Carbone
LARM: Laboratory of Robotics and Mechatronics, University of Cassino, Italy

Asma Merdassi, Laurent Gerbaud and Seddik Bacha
Grenoble INP/ Grenoble Electrical Engineering Laboratory (G2Elab) ENSE3, Domaine Universitaire, France

Esteban Chávez Conde
Universidad del Papaloapan, Campus Loma Bonita, Mexico

Francisco Beltrán Carbajal
Universidad Autónoma Metropolitana, Unidad Azcapotzalco, Departamento de Energía, Mexico

Antonio Valderrábano González and Ramón Chávez Bracamontes
Universidad Politécnica de la Zona Metropolitana de Guadalajara, Mexico

Christophe Versèle, Olivier Deblecker and Jacques Lobry
Electrical Engineering Department, University of Mons, Belgium

Khalaf Salloum Gaeid and Hew Wooi Ping
University of Malaya, Malaysia

S. M. Mizanoor Rahman and Haoyong Yu
National University of Singapore, Singapore

Ryojun Ikeura
Mie University, Japan

Index

Printed in the USA
CPSIA information can be obtained
at www.ICGtesting.com
JSHW051349091023
49903JS00006B/81